PHILIP'S

Little Book of
ECLIPSES

An Essential
Companion to
Solar & Lunar
Events

TOM KERSS & DR RYAN FRENCH

Tom Kerss is the author of a dozen books for readers of all ages, including educational fiction for young stargazers and coffee-table titles spanning all manner of cosmic curiosities. He is a veteran sky-chaser working with eclipses, the Northern Lights and dark sky sites around the world. Tom is a regular media expert with hundreds of broadcast interviews, documentary items and print contributions. He holds a degree in astrophysics, a masters in spacecraft engineering, and was formerly based at the Royal Observatory in Greenwich. He splits his time between the UK and the US.

Dr Ryan French is a solar physicist, science communicator and author. He is pursuing the mysteries of the Sun at the forefront of modern solar physics research, using cutting-edge telescopes on the ground and in space. His research takes him all over the world, collaborating with the global community of solar physicists. Ryan also works to share the wonders of the Sun and space with the public through social media, public talks and occasionally on television and radio.

First published in Great Britain in 2026 by Philip's, an imprint of
Octopus Publishing Group Ltd
Carmelite House
50 Victoria Embankment
London EC4Y 0DZ
www.octopusbooks.co.uk

An Hachette UK Company
www.hachette.co.uk

The authorized representative in the EEA is Hachette Ireland, 8 Castlecourt Centre,
Dublin 15, D15 XTP3, Ireland (email: info@hbgi.ie)

Text copyright © Tom Kerss and Ryan French 2026
Design and layout copyright © Octopus Publishing Group 2026

Distributed in the US by Hachette Book Group
1290 Avenue of the Americas, 4th and 5th Floors
New York, NY 10104

Distributed in Canada by Canadian Manda Group
664 Annette St., Toronto, Ontario, Canada M6S 2C8

ISBN: 9781849077545
eISBN: 9781849077538

A CIP catalogue record for this book is available from the British Library.

Printed in Dubai.

10 9 8 7 6 5 4 3 2 1

CONTENTS

INTRODUCTION

I N THE YEAR 528, a bewildered man was dragged in front of a royal court, suspected of being a sorcerer. The evidence? His unusual clothing and manner of speech. At the behest of his jealous mystic, the King passed summary judgement and sentenced the ill-fated stranger to death. Just before his execution (by burning at the stake) commenced on 21 June, a total solar eclipse brought darkness to the castle courtyard. The doomed man, one Hank Morgan, had foreknowledge that this would happen, and he used it to convince the King to spare his life. Indeed, by returning the Sun he gained great power and influence across the kingdom. How did Morgan know the eclipse would occur? He had travelled from the future.

So wrote Mark Twain in his 1889 novel *A Connecticut Yankee in King Arthur's Court*, in which an American engineer, who wakes in Arthurian England after a blow to the head, outwits the fraudulent magician Merlin to become King Arthur's right-hand man. As it happens, no eclipse occurred on that date, but it's not implausible that a person in Morgan's position could win favour by predicting one. The story itself was probably inspired by true events, which occurred almost four centuries before Twain wrote it (and almost a millennium after it was set). In 1504, stranded for more than six months on the island now known as Jamaica, Christopher Columbus faced straining tensions with the indigenous Arawak people. Armed with an almanac, he spun a story to the natives, declaring that if they stopped provisioning the crew, God would show his displeasure and make the moonrise of 29 February appear, to quote his son Ferdinand, 'inflamed with wrath'. The total lunar eclipse Columbus knew would occur struck fear into the hearts of the Arawak, who vowed to continue helping him. Carefully timing

This book covers both solar and lunar eclipses, each of which offer unique experiences that instil awe and wonder.

the eclipse from in his tent, he finally announced God's forgiveness shortly before it subsided, and as a result, he and his crew survived until their rescue five months later.

It's hard to imagine anyone being duped by this kind of gambit today, and that's a testament to our centuries-long march of science and education. As our collective understanding of eclipses has grown, so too has our appreciation. Events that once fuelled superstition or instilled terror now feed wonder and invite admiration. We don't cower at the unexpected behaviour of the Moon or Sun. We calculate it, predict it and even chase it. Lunar and solar events offer opportunities to be thrilled at home and abroad; to contemplate; to explore our own existence as small but significant observers of a grand celestial clockwork. Granted, much of the charming mystery of eclipses has been swept away over time, but their awesome power to astound is untarnished. Our experiences are enhanced by the inherited knowledge of generations of astronomers. There are intricacies and subtleties that you can look for yourself, which reveal unseen details about our home planet and its vital star. There are also new ways to witness, by eye or using a camera, and it's never been easier to produce your own stunning images.

So it is that we, the authors, offer you this guide with the hope that it will inspire you to seek out and make the most of the eclipse opportunities that come your way. We've filled it with our collective eclipse-chasing experience, as well as advice and curiosities, both historical and contemporary. May it serve you well. We wish you fair skies on your adventures!

‘ Lunar and solar events offer opportunities to . . . explore our own existence as small but significant observers of a grand celestial clockwork. ’

ECLIPSES
THROUGH HISTORY

WE ARE ALL creatures of the time in which we live, and though there are many universal aspects of the human experience that we share with our distant ancestors, it would be unwise to assume that we can fully comprehend how they perceived the natural world. In our age, we are able to predict the intricate dance of the Sun, Moon and stars thousands of years into the future with astonishing accuracy. We are so immersed in this body of wisdom that it has permeated our popular culture, and we are raised with a broad understanding of virtually every domain of science, even if we don't take a special interest in them. The chances are that you picked up this book with some prior knowledge of eclipses, and that you are awed by them rather than scared of them.

The list of inscrutable phenomena hasn't necessarily grown shorter since prehistory, but it has changed drastically. Quantum events and the evolution of galaxies – problems for today's best and brightest to investigate – didn't trouble the minds of ancient people. Rather, they focused on the world they could see, and its cyclical and recurrent nature. They found both providence and danger all around them, as seasons and stars arrived, departed and returned again. Occasionally, an unexpected occurrence broke the pattern, such as the sudden darkening of the Sun or the blood-red disfiguration of the Moon.

That these happenings captivated and terrified our forebearers from prehistory is a given. But for every rich tapestry of myths and beliefs, there is also a story of early scientific investigation that illuminates the struggle our ancestors faced in trying to understand their place in the cosmos. As time went by, the perseverance of human ingenuity prevailed then as it does today, and we see the understanding of eclipses transform. What was once unfathomable is now a veritable scientific tool; what was once terrible is now beautiful.

' In our age, we are able to predict the intricate dance of the Sun, Moon and stars thousands of years into the future with astonishing accuracy. '

COSMIC DRAMA IN RECORDED HISTORY

At Lascaux in France, cave paintings believed to be at least 17,000 years old appear to depict the Hyades and Pleiades star clusters in an early version of what would become the constellation Taurus, the Bull.

Ample archaeological evidence supports the assumption that prehistoric people were acutely preoccupied with the sky, though their interpretations of it during this long span of time were undoubtedly explained entirely through supernatural and religious frameworks. Numerous examples of rock art, cave paintings and burial sites are adorned with symbols that represent celestial objects, and in some cases we might be tempted to declare that comets, meteors, auroras and eclipses have been recorded. For early humans, living in a world where survival depended on understanding natural patterns, the appearance of a strange object or sudden failure of the Sun or Moon would have been profoundly disturbing. However,

Concentric markings carved into a standing stone in Cairn L could be a crude illustration of a solar eclipse in 3340 BCE.

impressions related to the sky may be more abstract than accurate, so we must be careful when speculating about them.

Once we reach the age of recorded history, the fog begins to clear. Ireland is home to possibly the earliest known attempt to document an eclipse. It has been suggested that a carving in stone at Loughcrew Cairn L, near Oldcastle in County Meath, known as Petroglyph C16, depicts a solar eclipse visible at the site in 3340 BCE. Irish archeoastronomer Paul Griffin was the first to tie the carving to the event which suggests that the island's native people have long been fascinated with the sky. Multiple experts have speculated that the carvings in Cairn L are of astronomical significance, and the concentric circles may have held particular significance for ritual practices. Charred remains of many individuals have been discovered directly in front of the carvings – perhaps Neolithic priests interpreted the eclipse as a blessing for the location.

Before standing stones were used as potential canvases for eclipse carvings, Chinese court astronomers kept meticulous records of eclipse observations as early as the 2nd millennium BCE. Chinese mythology, as with so much of the natural world, contextualized eclipses through the actions of dragons – in this case, the cosmic drama of celestial dragons devouring the Sun

or Moon. From this superstition grew a sophisticated system of political determinism, in which eclipses served as omens affecting imperial legitimacy. The emperor, as the Son of Heaven, was responsible for cosmic harmony, and eclipses indicated a disturbance in the natural order that required ritual intervention. During eclipses, astronomers beat drums and gongs to frighten away the dragons, a practice that persisted for millennia.

In Mesopotamia, the Babylonians (2000–1600 BCE) developed some of the most advanced early astronomical knowledge alongside their rich mythological framework. Arguably, their records of celestial events set the highest standard of the era and are still used in research today. The Babylonians associated eclipses with the gods' displeasure, viewing them as portents of political upheaval, warfare or the deaths of kings. The Babylonian creation myth, the Enuma Elish, describes cosmic battles between primordial forces, and eclipses were considered to be echoes of these eternal conflicts. Yet within the constraints of these myths, Babylonian astronomers began to develop prototype mathematical models to predict eclipses, resulting in the earliest known form of celestial mechanics.

EGYPTIAN AND NEAR EASTERN PERSPECTIVES: DEITIES IN PERIL

To a significant extent, ancient Egyptian civilization (3100–30 BCE) was centred on the worship of Ra, the Sun god, and his devout followers viewed solar eclipses as particularly ominous events. In Egyptian mythology, eclipses occurred when the serpent Apep temporarily succeeded in swallowing Ra during his daily journey across the sky. This tale of dread reflects deeper anxieties about cosmic order, since Ra's daily resurrection represented the triumph of life over death and order over chaos. Solar eclipses thus symbolized a temporary invasion of primordial chaos, requiring priests to perform protective rituals to ensure Ra's eventual return.

The Egyptians were captivated by skyward phenomena, but the records they left behind contain scant clues about

Travelling on his barge through the underworld (across the sky) Ra is protected by Set (a storm god), who spears Apep who is in serpent form. If Apep wins and consumes Ra, the Sun will be eclipsed.

↑

In Hindu astrology, Rahu and Ketu signify the lunar nodes, where the Moon crosses the ecliptic and can produce eclipses.

their understanding of eclipses. They did, however, leave a lasting legacy in astronomy. Observances of the sky played a role in their agricultural success, and they organized the calendar accordingly. They advanced a 365-day solar calendar anchored to the heliacal rising of the star Sirius, observed the phases of the Moon and maintained a religious lunar calendar. The Egyptians also used stars for timekeeping. However, the mathematics of eclipses and planetary circumstances seem to have been of little relative importance.

Similar patterns of deities in peril emerged throughout the ancient Near East (3000–330 BCE). Persian Zoroastrianism painted eclipses as battles between Ahura Mazda, the god of light, and Angra Mainyu, the destructive spirit of darkness. Hindu traditions spoke of the demons Rahu and Ketu, who periodically devoured the Sun and Moon as cosmic punishment.

The similarities between the various stories and rituals across the ancient world's great civilizations underscore a common reverence for the two most important objects in the sky – the Sun and Moon – along with the struggle to grapple with their occasionally inexplicable behaviour.

GREEK INNOVATION: FROM MYTH TO MATHEMATICS

The ancient Greeks gradually replaced purely mythological explanations in favour of the emerging field of natural philosophy. This crucial transition in eclipse thought was well documented. Early Greek poetry – particularly Homer's epics, *The Iliad* and *The Odyssey* (8th century BCE) – treated eclipses as divine interventions, often associated with the anger of the gods, or with warnings of impending disaster. The sudden descent into darkness (if not metaphorical, then brought about by a solar eclipse) appears in *The Odyssey* as an omen of doom for Penelope's suitors.

However, by the 6th century BCE, Greek philosophers were considering purely natural causes. Thales of Miletus (c. 624–c. 546 BCE), often regarded as the first Western philosopher, allegedly predicted a solar eclipse in 585 BCE

Solar eclipses are seldom good news in ancient stories. In Homer's *The Odyssey*, an eclipse signals the return of Odysseus and the slaying of the suitors. The seer Theoclymenus ominously proclaims, 'the Sun has been obliterated from the sky, and an unlucky darkness invades the world.' The event is depicted in this 1812 painting by Thomas Degeorge.

that occurred during a battle between the Lydians and Medes, causing both armies to cease fighting in terror. Whether Thales genuinely predicted this eclipse remains a subject of debate, but the story illustrates the growing Greek interest in understanding celestial mechanics through reason rather than mythology.

Philosopher Anaxagoras of Clazomenae (c. 500–428 BCE) insisted that eclipses resulted from the alignment of celestial bodies rather than divine intervention. He suggested that lunar eclipses occurred when the Earth cast its shadow on the Moon, and solar eclipses occurred when

Did Thales of Miletus predict a solar eclipse? Greek historian Herodotus claimed he did, but his account was written about 150 years after the event. If the story is true, it is one of the most remarkable scientific predictions in the ancient world.

the Moon obscured the Sun. His insight, dating to around 430 BCE, was far reaching. Indeed, it is precisely correct. Fortunately, it did not go unnoticed.

The scientist and philosopher Aristotle (384–322 BCE) was influenced by Anaxagoras, further developing his forerunner's ideas and using eclipse observations to argue for the Earth's spherical shape. He noted that its shadow on the Moon during lunar eclipses was always circular, regardless of timing, and therefore it demonstrated that Earth must be a sphere. It is an example of the sophisticated scientific inference for which the Greeks ultimately became known, and it illustrated the potential of indirect observation, which is commonplace in astronomy.

Greek astronomers were fascinated by eclipses, which proved invaluable to their understanding of the wider cosmos. Here, a partial lunar eclipse can be seen over the Acropolis in Athens.

HELLENISTIC PRECISION

A romanticized Victorian impression of Hipparchus observing the stars at the observatory in Alexandria. There is no evidence he visited the city, but he did correspond with astronomers there. Later, Ptolemy lived and worked in the city while using data compiled by Hipparchus.

The Hellenistic period (323–31 BCE) saw the culmination of ancient progress in the works of astronomers such as Hipparchus and Ptolemy. In the 2nd century BCE, Hipparchus of Nicaea (c.190–120 BCE) developed one of antiquity's most sophisticated eclipse prediction systems. He combined meticulous astronomical observations with early mathematical analysis to discover the fundamental periodic nature that governs lunar and solar eclipses. At the heart of his system was a cycle occurring on a period of approximately 18 years, 11 days and 8 hours (or 223 synodic months) during which the Sun, Moon and Earth return to nearly identical geometric configurations. This is now known as the Saros cycle (see page 62).

Hipparchus recognized that his eclipse cycle occurred due to the fact that 223 lunar months (the period between two identical lunar phases, or 29.5 days) are almost exactly as long as 242 draconic months (the time taken for the Moon to return to the same node where its orbit intersects the ecliptic plane, or 27.5 days). This coincidence meant that eclipses would recur at the same lunar nodes with similar characteristics. By carefully cataloguing eclipse observations from centuries of excellent Babylonian records, Hipparchus identified eclipses belonging to the same Saros series, allowing him to predict when future eclipses would occur by simply adding the Saros period to known eclipse dates. His system also incorporated the 19-year-long (235 lunar-month-long) Metonic cycle, which repeats when lunar phases recur at the same time of year. It was by far the most advanced approach to date, and it revealed the Moon's complex orbital variations as never before. Hipparchus developed geometric models to calculate the timing, duration and magnitude of upcoming eclipses with remarkable precision. His approach represented a revolution in astronomy, connecting pure observation with mathematical precision. It allowed Hipparchus to describe eclipses decades or even centuries in advance, with an accuracy that wouldn't be significantly improved until the development of modern orbital mechanics in the early 17th century, and it established him as one of history's greatest computational astronomers.

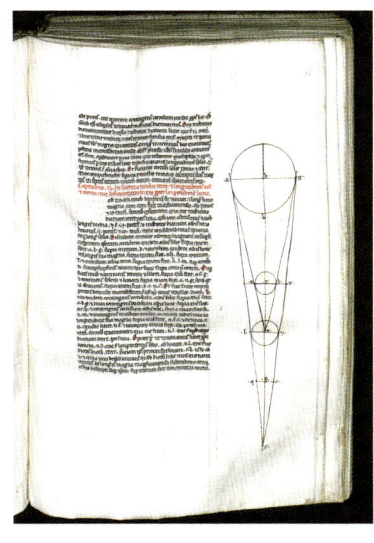

A page from Ptolemy's *Almagest* describing the circumstances that give rise to lunar and solar eclipses.

During the 2nd century, the astronomer and geographer Ptolemy (c. 100–170) produced his *Almagest* (*Great Work*). It developed more precise mathematical models and world-class, comprehensive tables which significantly refined and systematized Hipparchus's eclipse prediction method. Ptolemy was able to account for subtle variations in the Moon's orbit that Hipparchus had observed but had not been able to fully explain, such as its eccentricity. Ptolemy created detailed lunar and solar tables that incorporated its irregularity to account for changes in the Moon's orbital speed. His mathematical framework was constrained by his belief in a geocentric cosmos, with the Earth at the centre of everything, and his geometric models relied on epicycles and deferent circles to represent the Moon's non-uniform motion. However, despite being fundamentally flawed, they proved very capable of predicting not only when eclipses would occur, but also from where they would be visible.

Most importantly, Ptolemy's *Procheiroi Kanones* (*Handy Tables*) provided generations of working astronomers with ready-made computational tools that eliminated the need for complex geometric calculations, making eclipse prediction accessible across the ancient world. His systematic approach was so comprehensive and accurate that it remained the standard method for eclipse prediction throughout the Islamic Golden Age (8th to 13th centuries) and medieval Europe (5th to 15th centuries). Astronomers continued to use periodically refined versions of Ptolemy's tables until the advent of Newtonian mechanics in 1687.

 Astronomers continued to use periodically refined versions of Ptolemy's tables until the advent of Newtonian mechanics in 1687.

There is perhaps no more remarkable a demonstration of the tenacity of ancient eclipse investigators than the Antikythera mechanism, a bronze clockwork device discovered in a shipwreck off the Greek island of Antikythera in 1901. Dating to around 150–70 BCE, and probably influenced by the work of Hipparchus, this extraordinary artefact has been hailed as the world's first known analogue computer. Among its numerous functions, it could predict solar and lunar eclipses with impressive accuracy,

→

Ptolemy's geocentric model required continuous adjustment to fit his observations. It became increasingly complex and inelegant before its popularity waned.

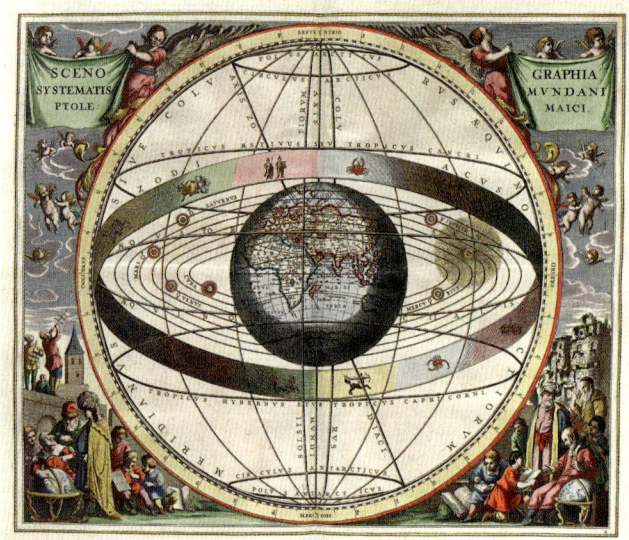

The remains of the Antikythera mechanism have fascinated both historians and astronomers alike.

incorporating both the Saros and Metonic cycles that govern eclipse patterns.

The device is an orrery which contains at least 37 meshing bronze gears that model the movements of the Sun and Moon (and possibly the planets) with a degree of precision comparable to mechanical clocks made in medieval Europe. Technologically, it is ahead of its time by more than a millennium. Its capabilities include not only the prediction of eclipses, but other eclipse circumstances, such as the time of day they will occur and the degree of obscuration.

The Antikythera mechanism is a genuine marvel, representing the vast intellectual distance travelled by our ancestors between the earliest days of recorded history and the Hellenistic period. The ancient world's eclipse legacy profoundly influenced subsequent scientific development. Islamic astronomers continued to build upon Greek foundations, and medieval European scholars, in a somewhat roundabout process, eventually rediscovered their ancient insights, setting the stage for the next scientific revolution.

This reconstruction of the Antikythera mechanism shows how the original may have appeared. In its original state, it was a masterpiece of engineering, capable of estimating the positions of Solar System objects far into the future.

NATURE OF THE SOLAR SYSTEM AND BEYOND

↑

The 1919 total solar eclipse proved Einstein's theory of general relativity.

Solar and lunar eclipses have long shaped the way we think about the world. In ancient Greece, the round shadow cast onto the Moon during a lunar eclipse revealed to ancient astronomers that the Earth was spherical, not flat. Measuring the timing of solar eclipses also enabled the Greeks to estimate the size and distance to the Sun, providing the first measurements of the scale of the Solar System. These contributed to the proposal by Aristarchus of Samos (310–230 BCE), a Greek astronomer and mathematician, that the Sun, not the Earth, was the centre of the Solar System. Unfortunately, this heliocentric viewpoint was not widely accepted in ancient Greece, and the geocentric model of the Solar System was favoured until the work of Polish astronomer Nicolaus Copernicus (1473–1543) in the 16th century.

Even in the 21st century, ancient eclipse observations can teach us about our place in the Solar System. In Anyang, ancient China, astronomers recorded measurements of the precise timing of total solar eclipses in 1226, 1198, 1172, 1163 and 1161 BCE. Using these, modern-day NASA scientists have measured minute changes in the speed of Earth's rotation during the past three millennia.

Beyond our Solar System, total solar eclipses have taught us about the fundamental nature of our universe, via the proof of Einstein's theory of general relativity, which was published in 1915. The theory postulated that gravity was not a force, but a consequence of the curvature of space-time in the universe. Imagine placing a heavy bowling ball in the centre of a large trampoline. The bowling ball creates a dip in the trampoline, causing small objects to roll towards it. This is the principle of general relativity, stating that massive objects in space (like the Sun, stars, black holes and so on) distort space, and cause other objects to feel a gravitational pull towards them. One measurable prediction of Einstein's theory was that the path of light should appear to bend around massive objects – such as the Sun.

Einstein's initial theory of relativity was published in 1911. Astronomers then sought to prove it observationally during a total solar eclipse. The first attempt, in 1914, was thwarted by the outbreak of World War I. This was fortunate for Einstein, who went on to publish a revised and corrected version of his theory four years later. Observations of the 1914 eclipse would have proven Einstein wrong.

The first opportunity to validate Einstein's new (and correct) theory came during the 1919 total solar eclipse. British astronomers Arthur Eddington (1882–1944) and Frank Dyson (1868–1939) travelled to Príncipe, an island off the west coast of Africa, to conduct what became known as the 'Eddington experiment'. Their aim was to use the darkness created when the Moon blocked the Sun to accurately measure the position of stars close to and behind the Sun. General relativity predicted that the apparent location of the stars would differ slightly from their known position, as the Sun's gravity bends the incoming starlight and distorts their location. Eddington and Dyson measured the background stars and found that they differed from their expected position by 0.0002 degrees, exactly as predicted. In proving Einstein's theory of general relativity by observation, the 1919 eclipse became the most important of all time.

> ❝ In proving Einstein's theory of general relativity by observation, the 1919 eclipse became the most important of all time. ❞

EXPEDITIONS TO UNDERSTAND THE SUN

The first organized eclipse expedition is often credited to Harvard University professor Samuel Williams (1743–1817). In 1780, he led a group of astronomers from the university in Massachusetts to Maine to view the event. Upon arrival, they found the area was occupied by British soldiers fighting the Revolutionary War. However, hope was not lost, as a mutual agreement between governments allowed Williams to pass into enemy territory to record the event...or so he thought. Unfortunately, Williams had miscalculated the eclipse path, and missed the total solar eclipse by a few miles.

FIRST PHOTOGRAPH OF A TOTAL SOLAR ECLIPSE

The first photograph of a total solar eclipse was taken on 28 July 1851, at the Royal Observatory in Königsberg, Prussia (now Kaliningrad, Russia). The image (below left) clearly captures the solar corona. Although cameras were available at the time, the technology was not widely accessible, and the recording of solar eclipses continued primarily through sketches rather than photographs for a few more decades.

The first photograph of a solar eclipse, taken in 1851.

A typical sketch of the solar corona during a total solar eclipse, drawn in India during the 1898 eclipse.

During the total solar eclipse of 1806, Spanish astronomer José Joaquín de Ferrer (1763–1818) travelled to upstate New York, USA, to view the event (see page 86). He studied the faint white glow visible during a solar eclipse, and proved it to be caused by the Sun and not the Moon (as previously thought). He had discovered, for the first time, that the Sun (and therefore other stars) has an atmosphere. He named the Sun's atmosphere the solar corona, after the Spanish word for crown.

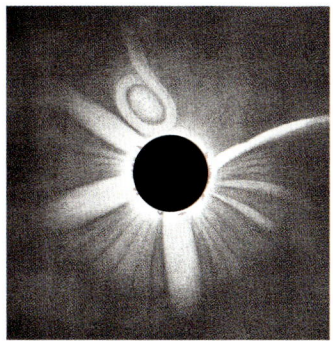

The first observation of a coronal mass ejection, sketched during the 1860 total solar eclipse.

In 1860, another Spanish astronomer Guglielmo Tempel sketched a view of a total solar eclipse. Unlike typical eclipse sketches, which capture bright streaks emanating away from the Sun within the solar corona, Tempel noted and sketched a twisted spiral-like structure towards the bottom-right of the corona. It was not known at the time, but Tempel had likely recorded the first ever observation of a coronal mass ejection (CME, see page 96). CMEs are the eruptions of plasma from the Sun's atmosphere, and responsible for the strongest aurora experienced here at Earth. Despite Tempel's observations, the existence of coronal mass ejections would not be verified until the second half of the 20th century.

The next major solar eclipse breakthrough came in 1868, with the discovery of a new element. In chemistry, different elements emit and absorb unique wavelengths of light, which differ like a barcode or fingerprint. This is a fundamental principle of spectroscopy, the measurement of the spectrum of light at different wavelengths. During the 1868 total solar eclipse, British astronomer Norman Lockyer (1836–1920) measured the emission of an odd wavelength of light from the solar corona. Its existence did not correspond to any known elements, and so Lockyer had discovered a new element within the solar corona. Believing that this newly found element was unique to the Sun, he named it after the Greek word for the Sun, 'Helios'. Lockyer had just

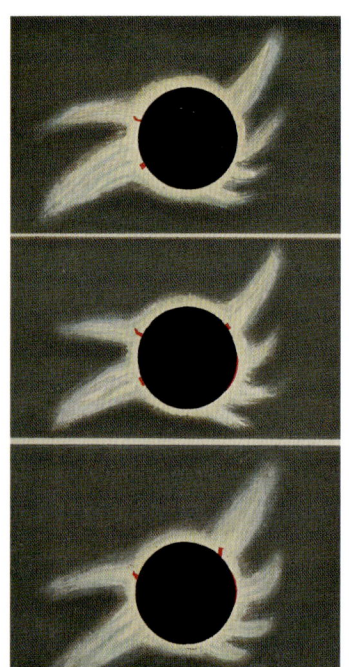

discovered and named helium, the second most abundant element in the universe (after hydrogen) that makes up 26 per cent of the Sun. Despite this, it had never been discovered before and would not be isolated on Earth until 1882 when Italian scientist, Luigi Palmieri (1807–96) identified it in gases released by the Mount Vesuvius volcano.

In 1869, astronomers Charles Young (1834–1908) and William Harkness (1837–1903) hoped to repeat Lockyer's success by discovering their own new element during that year's total solar eclipse. Independently of one another, they found an unknown wavelength of light emitted in the green part of the solar spectrum. Believing they had discovered a new element, they named their new discovery 'Coronium' after the solar corona. Unfortunately for them, coronium turned out not to be a new element after all. Instead, the unknown wavelength of light they had discovered originated from a species of iron (Fe) with 12 electrons stripped away from its atom.

Sketches of the 1868 total solar eclipse, the event that led to the discovery of helium.

The visible light spectrum of the Sun, with black lines marking absorption from elements on the Sun.

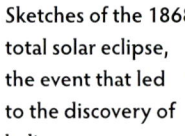

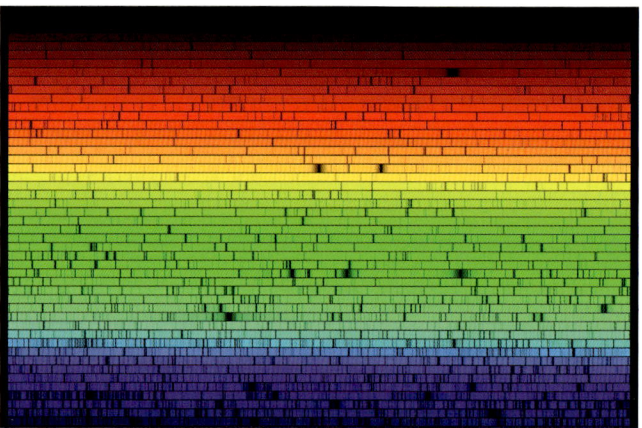

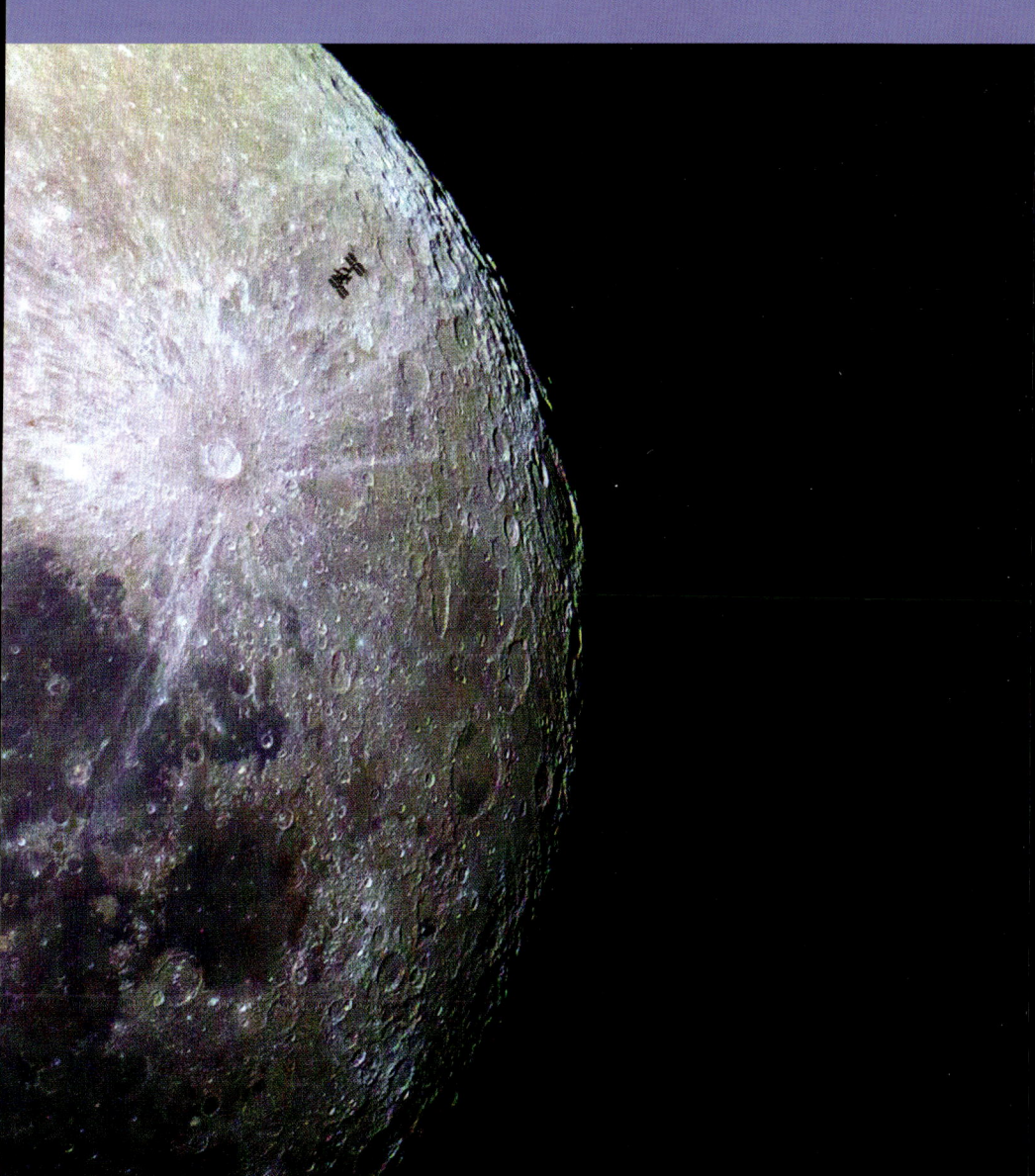

THE HISTORICAL TRIUMPH of science over mysticism delivered an extremely comprehensive body of knowledge. But science doesn't end. Today, curiosities remain, and eclipse science has branched out to impact other domains of astronomy. In this chapter, we will look at some recent and ongoing areas of research, as well as the ways eclipses affect modern science and technology.

THE THERMAL CHARACTERISTICS OF THE MOON

Can you recognize the face of our Moon? The major features are visible in this infrared image from MSX.

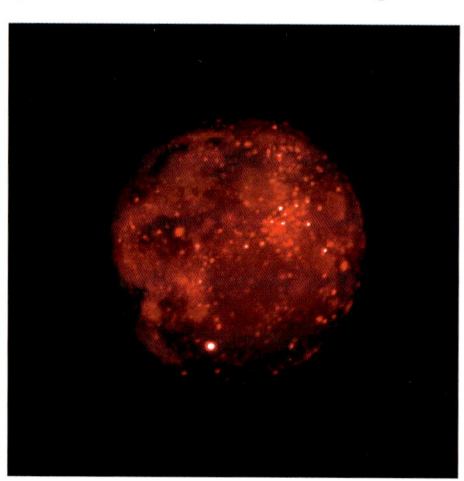

With no atmosphere to lock in heat, the Moon's nightside is substantially cooler than its dayside. When nighttime comes, there is a dramatic thermal shift. The searing heat of direct sunlight, which is around 120°C (250°F) becomes a frigid cold, plummeting to approximately -130°C (-200°F). This extreme temperature swing causes the lunar surface to contract, creating seismic waves that can be detected by sensitive instruments. The thermal stress accentuates fractures, kicks up dust particles and generates what scientists call thermal moonquakes – subtle but measurable tremors that ripple under the airless surface.

The properties of these moonquakes (and others caused by gravitational forces) have been investigated in some detail thanks to the Apollo 11, 12, 14, 15 and 16 missions (1969–72), when seismometers were deployed on the Moon's surface.

This seismic experiment deployed on the lunar surface by Buzz Aldrin during the 1969 Apollo 11 mission is one of five on the Moon. Tests on subsequent missions also delivered valuable data. The Artemis programme will deliver improved lunar seismology.

Unfortunately, the accuracy and reliability of the data is quite poor, but there is some speculation that sufficiently capable instruments could investigate the possibility of thermal stress, when the Moon's surface is allowed to cool relatively suddenly during a lunar eclipse. One of the key scientific goals of the Artemis programme, which began in 2017, is to improve lunar seismology, potentially providing enough sensitivity to probe the effects of lunar eclipses, and in so doing come to new conclusions about the interior of the Moon.

On 27 September 1996, the Midcourse Space Experiment (MSX), principally designed to detect ballistic missiles, used its SPIRIT-III infrared imager to capture a total lunar eclipse. The instrument enabled scientists to characterize the thermal distribution of the surface in the umbral region of the Earth's shadow, and the image reveals that different regions cool at surprisingly different rates. The lunar maria (seas) are identifiable, retaining heat for longer and thus appearing brighter than the surrounding highlands. Tycho, a prominent crater, makes a particularly dazzling spot, as it is still considerably warmer than most of the surface. The rate of cooling has implications for the way craters deform over time and the production of lunar soil or regolith. The impacts of eclipses on these processes are not yet known.

SCIENCE OF THE SOLAR ATMOSPHERE

Total solar eclipses remain highly valuable for scientists researching the Sun. They provide views of the solar atmosphere otherwise impossible to obtain from the ground. Although it is possible to send telescopes to space to observe the solar atmosphere (see page 33), this is expensive, and it can be decades between the initial mission concept to launch and data collection. Total solar eclipses offer the opportunity for scientists to use lower-budget telescopes to research the solar corona from the ground. But there are two primary limitations. The first is that total solar eclipses can happen anywhere, often in remote locations. This can provide logistical challenges when transporting equipment and scientists to the eclipse sites. The second is that eclipses are brief and infrequent. On average a total solar eclipse happens once every one and a half years. When one does happen, the solar corona is only visible for a few minutes – providing a limited window during which scientific data can be collected. The longest duration of a total solar eclipse is a little over 7 minutes, but most are far shorter than this, ranging from 1 to 4 minutes. Despite these limitations, each total solar eclipse sees a flurry of scientists from around the world travelling to the best site, bringing with them a range of exciting scientific instruments to research the solar corona.

There are too many ground-based eclipse experiments to mention them all, but the techniques discussed here use creative methods to collect more than a few minutes of data during a total solar eclipse.

A single location along the central eclipse path will witness the solar corona for a few minutes, but it can take several hours for the total eclipse to travel across the Earth's surface (see page

' On average a total solar eclipse happens once every one and a half years. '

This high-res processed image of the 8 April 2024 eclipse shows the Sun's corona in artificial colours that indicate the polarization or orientation of the light. Citizen scientists in Dallas collected these data through the SwRI-led CATE 2024 experiment.

55). In essence, if person A sees the solar corona at 10am, person B, a few thousand kilometres away, sees it at midday. Therefore, if you place a series of telescopes along the eclipse path, you can stitch together the data to obtain observations of the solar corona over multiple hours. This was the concept behind the Citizen CATE project. During the 8 April 2024 total solar eclipse in North America, scientists funded by the National Science Foundation and NASA placed 35 identical telescope setups along about 3,200 km (2,000 miles) of the approximately 14,500-km (9,000-mile) eclipse path from Texas to Maine. Data was collected by groups of community volunteers, culminating in several hours of observations of the solar corona. The telescopes measured the corona in polarized light, revealing information on the orientation of magnetic fields in the Sun's atmosphere. In the first image (above), released from the Citizen Continental America Telescope Eclipse (CATE) 2024 data by the Southwest Research Institute (SwRI) in San Antonio, Texas, the colours represent the angle of polarization in the Sun's atmosphere.

Another way to extend the duration of eclipse observations is to chase the shadow of the Moon. This isn't possible from the ground, but it can be achieved in the air. In 1973, portholes were created in the roof of Concorde 001's fuselage to allow eclipse observations from an altitude of 17.7 km (58,000 ft). The Concorde had five scientific instruments onboard, led by scientists from France, the United Kingdom and the USA. The aircraft, travelling at supersonic speeds of Mach 2, chased the shadow of the Moon over Africa, observing the solar corona for 74 minutes. This is the longest period of totality ever experienced.

Although Concorde was discontinued, high-altitude flights to chase eclipses continue. During the 8 April 2024 total solar eclipse in North America, independent eclipse experiments took place on two aircraft. A team led by SwRI flew a series of infrared cameras in the nose cone of a WB-57 jet, photographing the solar corona at previously unstudied wavelengths of light. Another experiment was led by the Harvard-Smithsonian Center for Astrophysics and National Center for Atmospheric Research. The Airborne Coronal Emission Surveyor (ACES) experiment was flown on a Gulfstream V plane to test instruments to measure the spectrum of emission in infrared light. Flying eclipse experiments at high altitudes (above 12.2 km/40,000 ft) has the advantage of measuring certain wavelengths of light in the corona, which otherwise are absorbed by Earth's atmosphere and not visible from the ground.

ACES experiment onboard the NSF NCAR Gulfstream V during the 8 April 2024 total solar eclipse.

↓

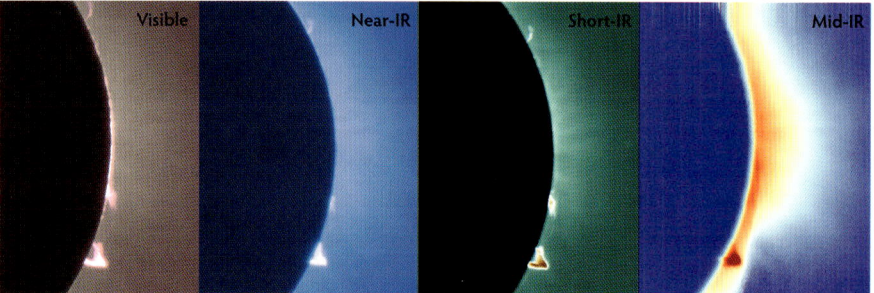

Above: The corona in four wavelength ranges, shown in preliminary images from a suite of visible-light and infrared imagers aboard a NASA WB-57 jet during the 8 April 2024 eclipse.

ARTIFICIAL ECLIPSES

While total solar eclipses are highly valuable for learning about the solar atmosphere, they are rare (every one and a half years). Fortunately, we can use technology to create 'artificial eclipses' to provide a more consistent view of the solar corona. These artificial eclipses are created using a type of telescope called a coronagraph. Much like a real solar eclipse, a corona-graph blocks light from the surface of the Sun, allowing longer exposure observations to capture faint details and structures in the solar corona. These telescopes can be used on the ground and in space. The majority use an occulting disk – essentially a small disk held up by a mounting arm in the centre of the tele-scope opening – to block incoming light from the Sun. Because sunlight can still scatter off the occulting disk, there is a limit to how close to the Sun's surface most coronagraphs can photo-graph the solar corona. With most telescopes, it is not possible to photograph the very bottom of the solar corona, but this can be easily seen during a total solar eclipse. New telescopes and technologies are being developed to counteract this light scattering problem.

Perhaps the most famous coronagraph is the Large Angle and Spectrometric Coronagraph (LASCO) onboard the NASA/ ESA Solar and Heliospheric Observatory (SOHO) satellite.

Coronagraph images of a CME from SOHO/LASCO C2 (*left*) and C3 (*right*). The faint grey circle marks the location of the Sun behind the occulting disk.

Coronagraph image from the NOAA CCOR-1 (*left*) and Proba-3's first published image of the solar corona (*right*).

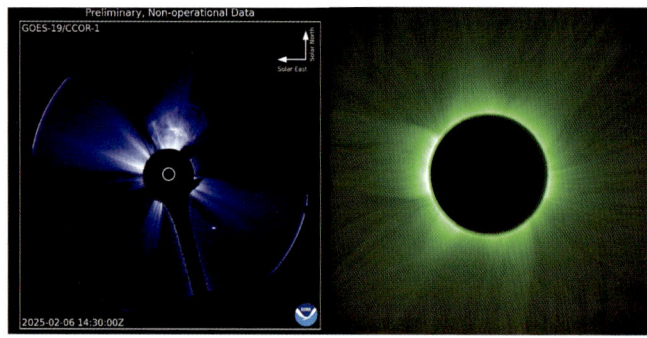

SOHO launched in 1995 with a fleet of instruments observing the Sun, but LASCO is the final operational instrument as of 2025. LASCO is comprised of two separate coronagraphs. The first, LASCO C2, images the solar corona from 0.5–6 solar radii (multiples of the Sun's radius) above the edge of the Sun. The second coronagraph, LASCO C3, has a wider field of view from 2.7 to 29 solar radii above the Sun's edge. LASCO has been a crucial resource in observing CMEs erupting off the Sun, and predicting their impacts at Earth. The effects of CMEs and other solar phenomena at Earth is called 'space weather'.

In 2024, the US National Oceanic and Atmospheric Administration (NOAA) launched the CCOR-1 coronagraph. CCOR is a more modern version of LASCO, and will eventually replace it in space weather operations. In recent years, coronagraphs have become progressively more advanced in their ability to observe the solar atmosphere. In 2025, the European Space Agency launched the Proba-3 mission. Rather than an occulting disk, it uses a second satellite, positioned 150 m (492 ft) away, to block out light from the Sun. Flying two satellites separated by this distance with enough precision for the second satellite to block out the Sun from the perspective of the first satellite, is an immense technological achievement. By using this method to block the Sun, Proba-3 can photograph much closer to the Sun's surface than regular occulting disk coronagraphs.

TRANSITS IN THE SOLAR SYSTEM AND BEYOND

When is an eclipse not an eclipse? Beyond solar and lunar eclipses, there are other eclipse-like phenomena in astronomy. They have different names, but the difference in definitions between them is not always clear cut. Let's introduce transits and occultations.

From the Earth's perspective, when the Moon passes in front of the Sun we experience a solar eclipse. If an apparently much smaller object passes in front of the Sun, we get a transit. In general, a transit is defined as a smaller object passing in front of a larger one. From Earth, we can view transits of the planets Venus and Mercury against the Sun. We only see transits of these two planets, as they are the only planets closer to the Sun than us – planets with larger orbits cannot pass in front of the Sun from our perspective. From further out in the Solar System than the Earth, other planetary transits can be visible, including that of the Earth in front of the Sun.

The transits of Mercury and Venus in front of the Sun are far rarer than total solar eclipses. They come in pairs separated by short gaps but with longer wait times between each pair. In some cases, one of the two transits within a pair will fall just above or below the Sun, allowing only one of the pair to be visible from the Earth.

For Mercury, transits (or near-misses) come in pairs with a three-year gap, separated from the next cycle of pairs by 7 years. The most

Comparison of a Venus transit (2012) and Mercury transit (2019), observed by the NASA Solar Dynamics Observatory.

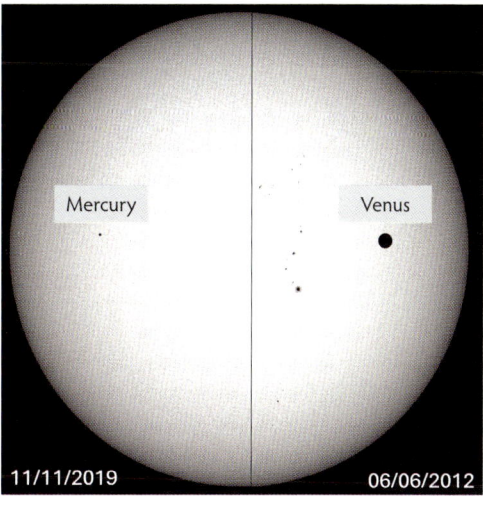

Mercury

Venus

11/11/2019

06/06/2012

recent transits of Mercury took place in 2016 and 2019. The next ten transits are:

1. 13 November 2032
2. 7 November 2039
3. 7 May 2049
4. 8–9 November 2052
5. 10–11 May 2062
6. 11 November 2065
7. 14 November 2078
8. 7 November 2085
9. 8–9 May 2095
10. 10 November 2098

Against the background Sun, Mercury is too small to see with a pair of solar viewing glasses (see page 79), and requires magnification from a solar telescope or similar (see page 83). Any techniques to observe solar eclipse with magnification (see page 84) can also be used to observe transits of Mercury.

Transits of Venus in front of the Sun are far more visually impressive than transits of Mercury. Because Venus is both larger and closer to Earth than Mercury, its apparent size against the background Sun is large enough to see it unmagnified with a pair of eclipse glasses. Venus transits are certainly impressive, but unfortunately you are most likely out of luck. The previous two transits of Venus were in 2004 and 2012, but the next two are not until 2117 and 2125.

Transits can occur with other objects too, including human-made ones. For example, the International Space Station, which is 109 m (357 ft)

A composite image of the International Space Station transiting the Sun.

↓

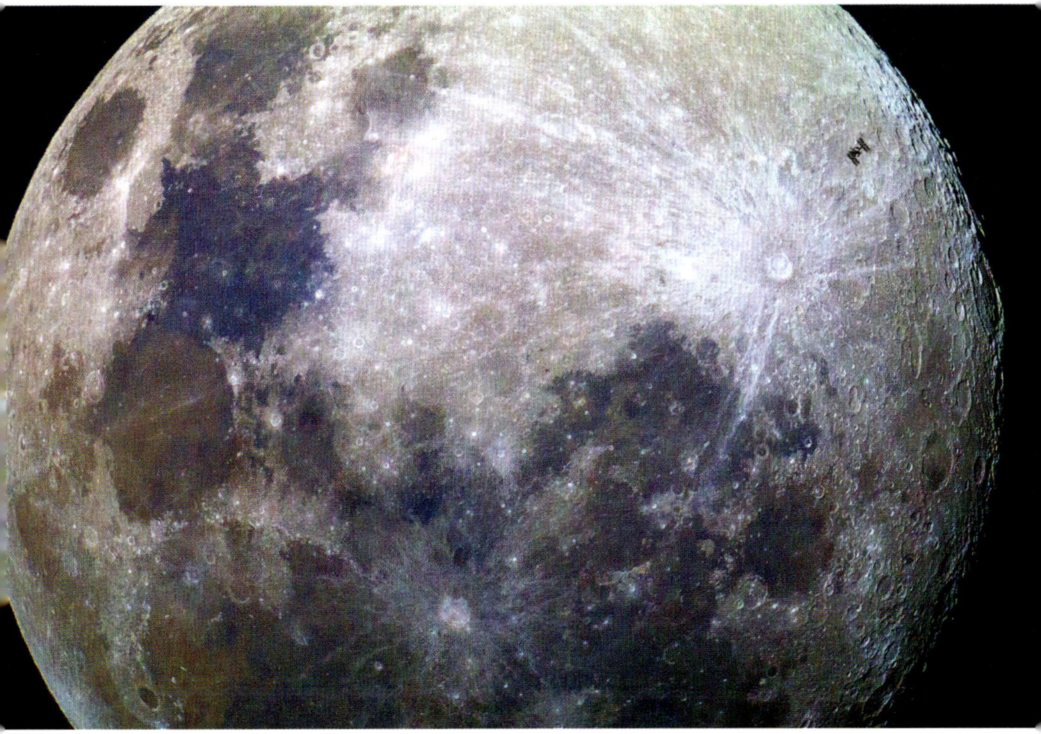

long and orbits the Earth at an altitude of 400 km (250 miles), frequently transits in front of both the Moon and the Sun.

If a transit happens as a smaller object passes in front of a larger one, and eclipses happen as an object blocks incoming light from another object, what is the difference between them? In essence, an eclipse is a special type of transit, meaning all eclipses are transits, but not all transits are eclipses. But at what point does a transit become an eclipse? Unfortunately, there is no satisfactory answer to this question. The simplest answer is that a transit becomes an eclipse when the blocking object becomes large enough, but this can be tricky to define with no specific threshold on which to base the definition. Another answer would be that an eclipse needs to cast a shadow, while a transit does not.

↑

The International Space Station, visible transiting over the upper-right region of the Moon.

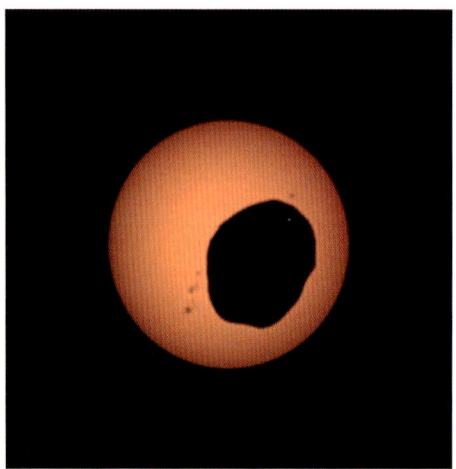

A solar eclipse on Mars, as the Martian moon Phobos transits the Sun.

Solar eclipses can happen on other planets in the Solar System, but look very different to solar eclipses on Earth. On Mars, the NASA Perseverance Rover has captured photographs of the Martian moon Phobos passing in front of the Sun. Because of the relative sizes and distances of the Sun and Phobos from Mars, the solar eclipse does not block the entire Sun, like our Moon does from Earth. For this reason, you could argue that this event is more of a transit than an eclipse. Although not as beautiful as solar eclipses seen from the Earth, observations of Martian eclipses can still reveal information about Mars. This includes providing the data needed to precisely measure the orbits of the Martian moons, and reveal insights into the structure of Mars's interior (which can change the moons' orbits).

We have also witnessed the shadows cast by solar eclipses on Jupiter, when any of the four Galilean moons (Io, Europa, Ganymede and Callisto) temporarily block sunlight for a small region of Jupiter's surface. Unlike Mars, where Phobos is too small to obstruct the Sun, from Jupiter's perspective its four largest moons appear much larger than the Sun and block out the sunlight entirely. Solar eclipses on Jupiter have been observed since the 17th century, and in 1678 were used by Dutch astronomer Christiaan Huygens (1629–95) to make an accurate measurement of the speed of light.

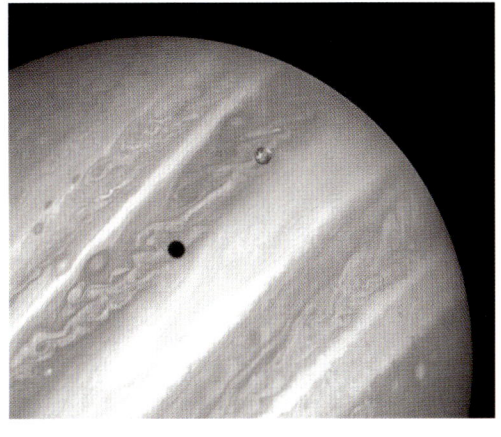

Transits can also teach us about planets far beyond our Solar System. The eight planets orbiting the Sun have been known since the discovery of Neptune in 1846. However, the

←↑

Solar eclipses on Jupiter, observed by the Hubble Space Telescope (*above*) and Juno (*left*) as Io casts a shadow on Jupiter's surface.

→

Occultation of Saturn behind the Moon.

first planet orbiting a star other than our own Sun was not recognized until 1995. These planets are called extrasolar planets, or exoplanets for short. By mid-2025, there were 5,926 confirmed exoplanets. The transit method is one of the three primary methods of detecting them. When the Moon passes in front of the Sun during a total solar eclipse, a small region of Earth experiences a 100 per cent reduction in sunlight. When Mercury or Venus pass in front of the Sun from Earth's perspective, or Phobos in front of the Sun from Mars's perspective, the Sun is not blocked completely, but the transiting Moon or planet creates a measurable drop in observed sunlight levels. This principle can be used to detect exoplanets around other stars. When observed, the brightness of a distant star is usually close to constant. However, when monitored over extended periods of time, some stars experience periodic consistently timed drops in brightness. These are caused by an exoplanet passing in front of the star from our perspective. In almost all cases, we cannot resolve the star or exoplanet directly, but the drop in light levels indicates that it's there. The timing and magnitude of the brightness drops can be used to determine the size of the exoplanet and its distance from its host star. As the exoplanet transits the star, astronomers also observe changes in the observed spectra, measuring elements within its atmosphere. This is just one way in which eclipses and transits teach us about objects beyond our Solar System.

Some stars experience periodic consistently timed drops in brightness. These are caused by an exoplanet passing in front of the star from our perspective.

Occultations (see page 42) are another eclipse-like phenomenon. They are the opposite of transits as they happen when apparently smaller objects pass in front of larger ones. For example, Mercury, Venus or any other planet passing behind the Moon (from our perspective) are considered occultations.

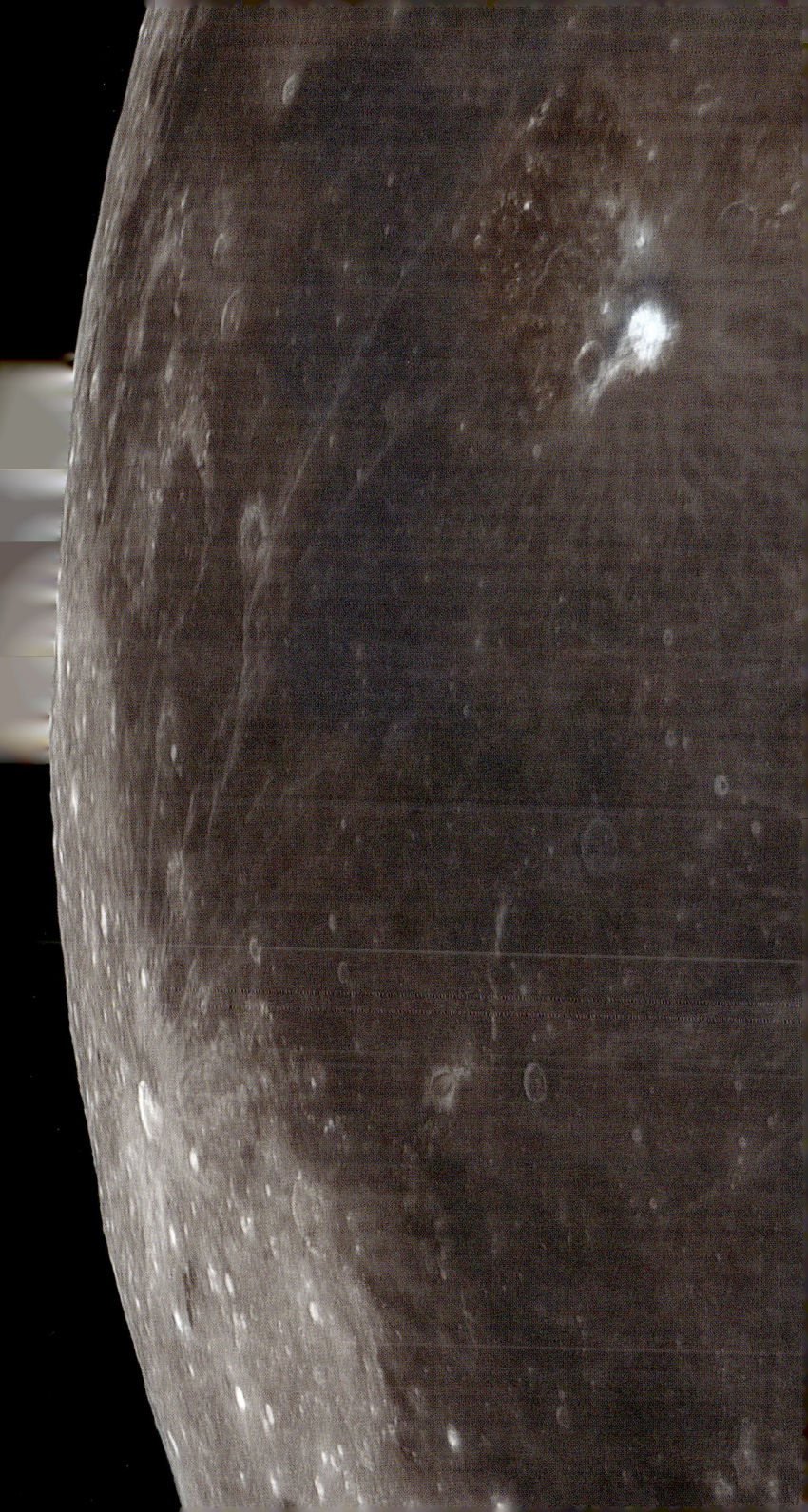

OCCULTATIONS: TINY BUT MIGHTY ECLIPSES

Not all eclipses involve the Sun or Moon. Astronomers use eclipses of other stars, called occultations, as a remarkable observational tool. When either the limb of the Moon or a dark asteroid passes in front of a star, the sharp edge of the occulting object acts like a knife, causing the starlight to disappear (and, in the case of an asteroid, perhaps reappear almost instantaneously). Very precise high-time resolution photometry of the star is used to measure its angular diameter, from which inferences about its true size can be made. This approach can also be used to detect unknown binary star systems, when the light curve shows complex patterns during the event. Grazing lunar occultations have enabled astronomers to make accurate determinations of the Moon's rugged terrain.

' *Grazing lunar occulations have enabled astronomers to make accurate determinations of the Moon's rugged terrain.* '

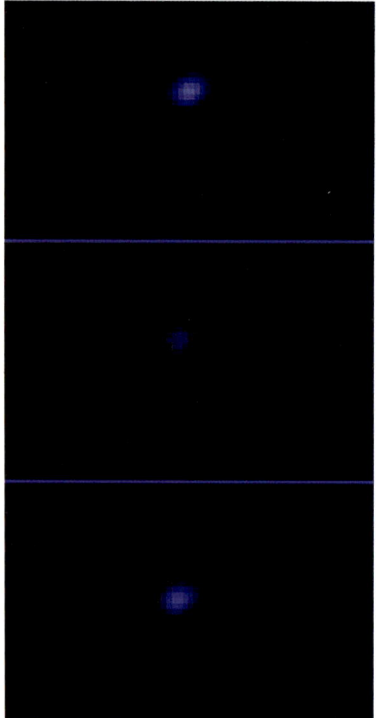

An occultation of the star UCAC4 345-180315 by the dwarf planet Pluto. The star is seen 5 minutes before the event (*top*), partially obscured during the event (*centre*) and 5 minutes after the event (*bottom*). The occultation allowed astronomers to study Pluto's tenuous atmosphere.

OCCULTATIONS IN THE KUIPER BELT

Occultations of asteroids and remote Kuiper belt objects are particularly valuable, as they reveal the exact size, shape and sometimes even the presence of moons around these distant icy or rocky bodies – information that would require expensive space missions to obtain otherwise. In 2014, occultations revealed a ring system around the asteroid Chariklo, a first-of-its-kind discovery.

The technique echoed the first conclusive detection of the rings of Uranus on 10 March 1977, when American astronomers James L. Elliot (1943–2011), Edward W. Dunham (1952–) and Jessica Mink (1951–) sought to measure the planet's atmosphere by the occultation of the star SAO 158687. Their observations, taken using the Kuiper Airborne Observatory (an infrared telescope mounted on a modified

An artist's impression shows how Chariklo's rings caused additional dips in the brightness of a background star during an occultation.

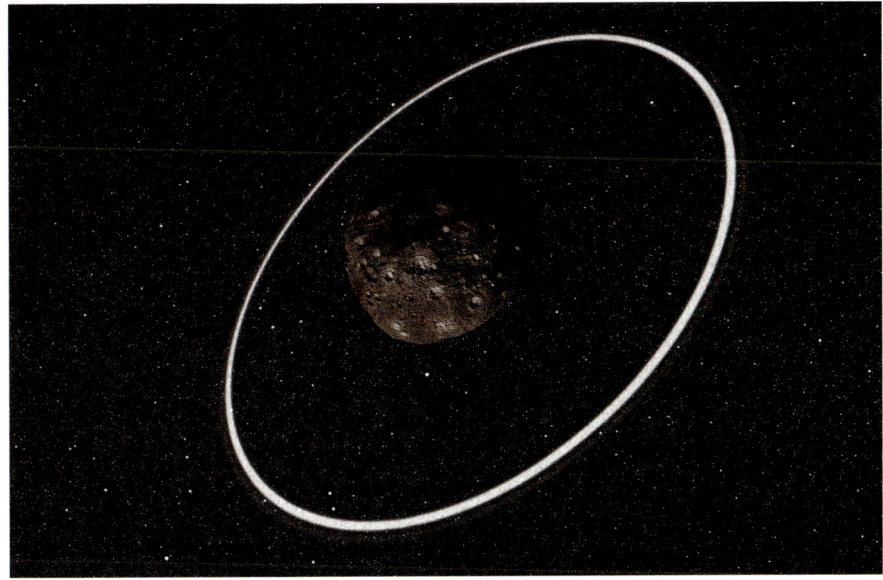

↑

Voyager 2 recorded an image of the rings of Uranus in 1986. They were first allegedly sighted by German-British astronomer William Herschel in 1789 and were officially discovered through occultation observations in 1977.

Pluto's hazy atmosphere was detected through a stellar occultation in 1988. In 2015, NASA's New Horizons spacecraft recorded this historic backlit image of the dwarf planet.

airplane), showed the star blinking in and out of view five times both before and after Uranus passed in front of it. This serendipitous discovery long preceded Voyager 2's arrival at Uranus in 1986.

Elliot continued to study occultations by other Solar System objects. In the late 1980s, again using the Kuiper Airborne Observatory, he conducted some of the first investigations of Pluto's recently discovered atmosphere, which had been detected during an occultation. Starlight was not sharply cut off by Pluto. Rather, its hazy envelope dimmed the light progressively. The precision with which occultations can be observed is continually improving with new generations of telescopes and detectors, and novel kinds of occultation-based science may be unlocked in the future.

THE EARTH'S ATMOSPHERE IN THE SHADOW OF THE MOON

Solar eclipses offer unique opportunities to study the Earth's atmospheric dynamics. The Moon's shadow sweeps across the planet at supersonic speeds exceeding 2,400 km per hour (1,500 miles per hour) causing rapid and measurable changes in atmospheric conditions. Temperature drops of 3–6°C (5.4–10.8°F) occur within the totality zone as solar heating ceases abruptly, but the cooling is delayed by up to 30 minutes because the atmosphere exhibits thermal inertia that is sensitive to humidity.

Wind patterns shift dramatically during totality, with the normal daytime convection currents weakening and

↑

Crew members from the Expedition 71 mission aboard the International Space Station witnessed the total solar eclipse on 8 April 2024. Here, the umbra of the Moon's shadow appears as a large, dark spot over Canada.

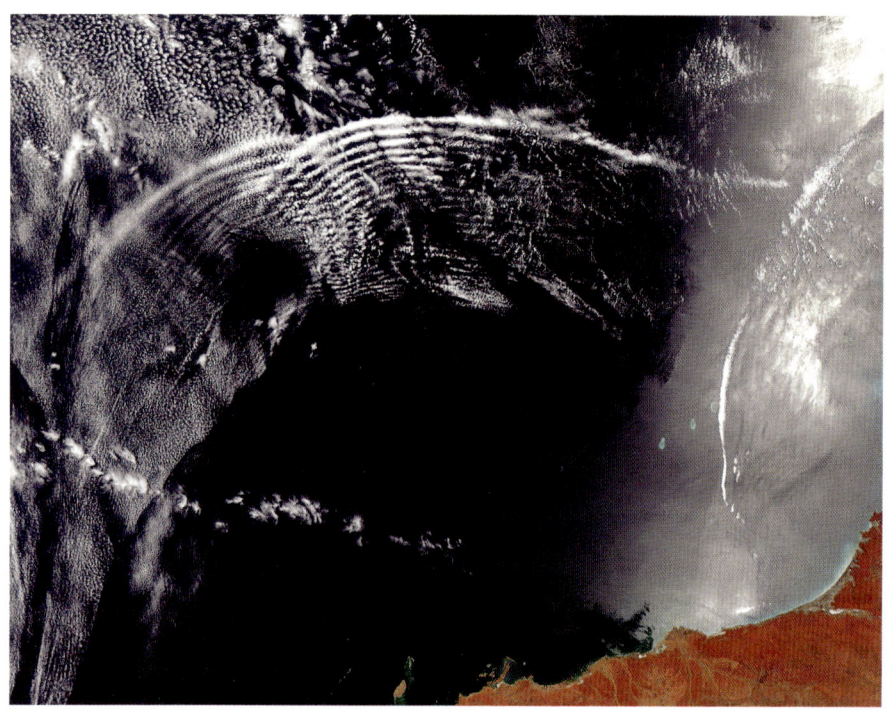

Atmospheric gravity waves, visible in this satellite image thanks to the cloud patterns they form, spread over the Indian Ocean. These waves emerge within the atmosphere where weather fronts are present, but eclipses have also been demonstrated to create them.

sometimes reversing direction, while humidity levels rise as the cooler air holds less moisture. The rapid cooling and subsequent rebounding of the air temperature can also generate atmospheric gravity waves – ripples in air density that spread outwards from the eclipse path and can be detected hundreds of kilometres away using sensitive barometric instruments.

Using radiosondes attached to weather balloons released every 15 minutes, researchers observed three potential stratospheric gravity waves during the total solar eclipse in Chile on 2 July 2019, with wave signatures appearing approximately 53, 62 and 156 minutes after totality. Surface pressure fluctuations of 10–30 pascals were recorded at ground level, generated by the rapid atmospheric cooling in both the stratosphere and troposphere as the eclipse shadow passed overhead. Four years later, students in New Mexico repeated the test, making the first conclusive measurements of eclipse-generated atmospheric gravity waves during the annual eclipse on 14 October 2023, validating predictions made more than 50 years earlier. Gravity waves are observable.

The ionized upper region of the atmosphere, called the ionosphere, responds dramatically to eclipse conditions, as the sudden reduction in extreme ultraviolet radiation from the Sun disrupts the ionized layers of Earth's atmosphere. Total Electron Content (TEC) measurements typically show decreases beginning shortly after the onset of significant obscuration, reaching minimum values about 15–30 minutes after the maximum eclipse (see page 57). Electron density in the so-called E-region (90–160 km or 56–100 miles in altitude) can decrease by up to 10 per cent during totality. These ionospheric changes cause temporary disruptions in radio communications and GPS signals, providing scientists with valuable insights into upper atmospheric structure and behaviour.

The total solar eclipse on 8 April 2024 recorded from a weather balloon similar to the ones used to measure atmospheric gravity waves.

DON'T FORGET THE ROBOTS!

Satellites experience significant and distinctly different challenges during lunar and solar eclipses, with each type of eclipse creating unique operational hazards as well as scientific opportunities. During lunar eclipses, satellites orbiting the Moon – such as NASA's Lunar Reconnaissance Orbiter (LRO) – undergo extreme thermal stress. As the Moon passes into the Earth's shadow, the spacecraft and others like it experience a sudden loss of solar heating, plunging into darkness along with the lunar surface. During eclipse seasons, which occur twice each year, LRO can be immersed in Earth's shadow for up to 48 minutes. These extended periods without sunlight render the spacecraft's solar panels ineffective, forcing a full reliance on battery power and exposing the satellite to rapid temperature swings – from the intense heat of direct sunlight to the deep cold of eclipse conditions within minutes.

Yutu-2, captured here by a camera on its lander Chang'e-4, survived the lunar eclipse on 16 May 2022.

The operational impact can be severe. Navigation systems become less reliable, and thermal management systems must work aggressively to protect sensitive instruments. For instance, in 2018, mission planners powered down LRO's inertial measurement unit to extend its operational life, reserving it for only the most critical scenarios, including lunar eclipses and safe mode entries. This highlights the demanding nature of eclipse conditions on lunar orbiters, where a brief eclipse can threaten both mission continuity and hardware integrity.

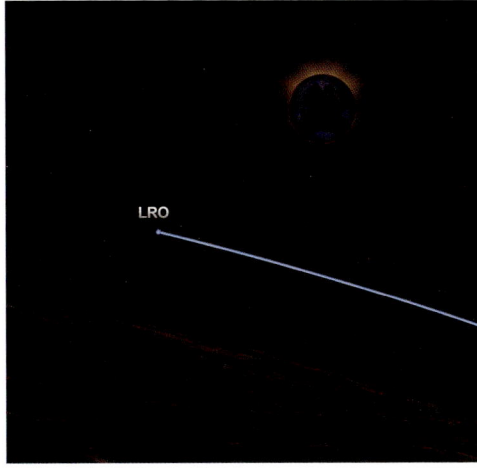

LRO

↑

An artist's impression shows the simulated view from the Moon of the Lunar Reconnaissance Orbiter in orbit close to the lunar surface during a total lunar eclipse.

Lunar landers and rovers are also vulnerable to the effects of lunar eclipses, particularly as they lack the relative flexibility afforded to orbiting satellites and must instead endure the full brunt of eclipse-induced environmental changes while stationary on the Moon's surface. The sudden drop and subsequent rise in temperature demand preparation. For example, during the lunar eclipse on 16 May 2022, China's Chang'e-4 lander and Yutu-2 rover, located on the Moon's far side, were powered down and placed in sleep mode to survive the extreme cold and preserve battery life. They had been sent there to carry out experiments and explore the geology. Surface temperatures during eclipses can fall below −130°C (−200°F), risking thermal contraction, battery degradation and even instrument failure if not properly managed.

In contrast, solar eclipses affect Earth-orbiting satellites in a very different way – primarily by their disruption of the ionosphere, a crucial region for radio wave transmission and

‘ *During eclipse seasons, which occur twice each year, LRO can be immersed in Earth's shadow for up to 48 minutes.* ’

Global Positioning System (GPS) satellites and other GNSS satellites experience eclipse seasons twice a year, degrading their precision.

satellite communications. The ionosphere, which extends from about 50 to 1,000 km (30 to 620 miles) above Earth, undergoes rapid changes during an eclipse as the sudden drop in solar extreme ultraviolet radiation leads to a significant reduction in ionization. As we saw previously, this results in temporary decreases in electron density, which can interfere with Global Navigation Satellite System (GNSS) accuracy and degrade satellite-to-ground communication links.

Navigational satellites themselves experience operational challenges during their biannual eclipse seasons, when they pass through Earth's shadow and temporarily lose direct sunlight. This loss causes solar sensors, which normally help maintain precise satellite orientation, to become ineffective. As a result, the satellites can exhibit unstable or less predictable yaw attitudes during the eclipse and for up to 30 minutes afterwards. Although GPS satellites use backup systems to control their attitude, these temporary instabilities can reduce the accuracy of their positioning signals. To ensure high-precision navigation, ground control stations often exclude data from satellites undergoing eclipse-induced attitude disturbances from critical GNSS calculations.

three

UNDERSTANDING ECLIPSES

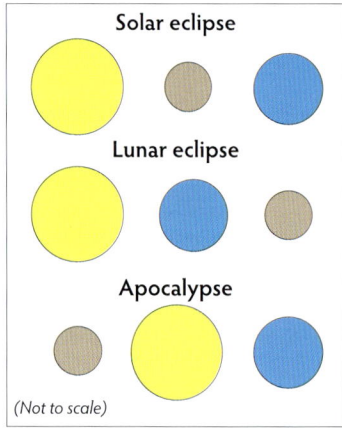

Solar eclipse

Lunar eclipse

Apocalypse

(Not to scale)

↑

The Sun, Earth and Moon arrangements needed for different eclipse types, including an old astronomy joke.

A LL ECLIPSES REQUIRE three objects. On the most basic level, they are defined as one object blocking incoming light from a second object and casting a shadow onto a third. From our vantage point on Earth, we can see two types of eclipses – solar and lunar. The three objects creating these eclipses are the Earth, Sun and Moon, and they are arranged differently for each type of eclipse. In essence, an eclipse is the alignment (or partial alignment) of these three objects in a straight line. For a solar eclipse, the three celestial bodies are ordered Sun → Moon → Earth. For a lunar eclipse, the order is Sun → Earth → Moon. The diagram (left) outlines the order of the Sun, Earth and Moon for solar and lunar eclipses, alongside a third, fictional scenario, used as a common astronomy joke.

ECLIPSES FROM THE EARTH

Earth experiences eclipses like no other planet in the Solar System. This is due to the cosmic coincidence of the relative sizes and distances of the Sun and the Earth from the Moon. The Sun has a diameter of 1.4 million km (864,000 miles), and is orbited by the Earth at an average distance of 150 million km (93 million miles). The Moon is significantly smaller with a diameter just under 3,500 km (2,160 miles), but is our closest neighbour in space orbiting at an average distance of 385,000 km (239,000 miles). So, what is the cosmic coincidence here? Well, the Sun is close to 400 times larger than the Moon, while also being 400 times further away. The primary result of this is that the Sun and Moon have the same apparent size in the sky. If you imagine the sky as a protractor, measuring 180 degrees from horizon to horizon, then the Sun and Moon each take up 0.5 degrees of the sky above us. This close similarity in apparent size between the Moon and Sun is what makes the range of eclipses we experience possible. There is no other location in the Solar System where a planet and moon form this fortunate link between their relative sizes and distances to the Sun.

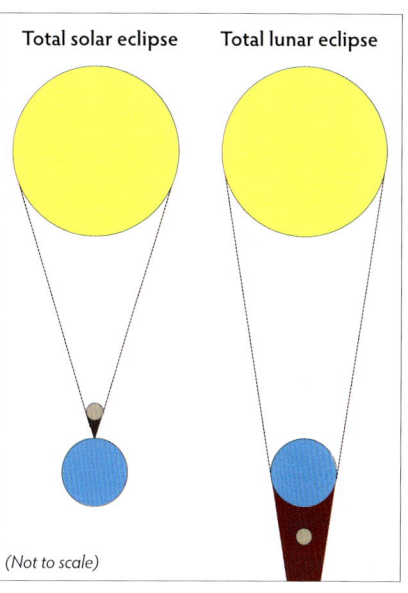

Total solar eclipse   Total lunar eclipse

(Not to scale)

↑

Schematic of shadows cast during solar and lunar eclipses.

The diagrams (right) outline how solar and lunar eclipses work. The sizes and distances between the Sun, Earth and Moon are not to scale, but the size-to-distance ratio remains constant for both the Sun and Earth. This allows us to accurately represent how shadows are cast during eclipses.

Solar and lunar eclipses can be summarized as follows:

→ **Solar eclipses** occur as the Moon passes directly between the Earth and Sun. As this happens, the Moon casts a shadow onto a region of the Earth and observers there witness some (or all) of the Sun being blocked by the Moon.

→ **Lunar eclipses** occur as the Moon passes into the shadow of the Earth, with the Earth blocking the majority of sunlight from reaching the Moon's surface. We witness this change in lunar brightness and colour as the lunar eclipse.

Both solar and lunar eclipses also have subcategories which are determined by specific differences in the precise alignments of the Sun, Earth and Moon (see table, page 54).

Eclipse category	Eclipse sub-category
Solar eclipse	Total solar eclipse
	Partial solar eclipse
	Annular solar eclipse
	Hybrid solar eclipse
Lunar eclipse	Total lunar eclipse
	Partial lunar eclipse
	Penumbral lunar eclipse

Sometimes the term 'central eclipse' is used. This is any eclipse in which the centre of the Moon's shadow (in a solar eclipse) or Earth's shadow (in a lunar eclipse) hits the Earth or Moon respectively. These include almost all total solar eclipses, annular solar eclipses and total lunar eclipses.

CATEGORIES OF SOLAR ECLIPSES

Solar eclipses are produced as the Moon passes between the Earth and the Sun. The exact nature of each one depends on two primary factors – the distance of the Moon from the Earth, and the relative angle between the Sun, Moon and Earth.

How does the distance between the Earth and the Moon influence a solar eclipse? The most impressive type of solar eclipse is the total solar eclipse. This happens when the Moon fully blocks the surface of the Sun from the perspective of a small region of the Earth. The size of the shadow cast by the Moon during a total eclipse is small, only a few hundred kilometres across. As the Earth rotates on its axis and the Moon and Sun move through their orbits, the location of the Moon's shadow moves. The motion of the Moon's shadow creates what we call the path of totality, with totality being the name of the eclipse phase where the Sun is fully blocked from view by the Moon. The shadow can take several hours to traverse the path of totality, meaning somebody standing at one end of the eclipse path will witness the total solar eclipse several hours before somebody at the other end, potentially halfway around the world. At any location, the totality phase lasts only a few minutes. The maps in chapter 7 (see pages 136–150) show solar eclipses, marking the path of totality as the Moon's shadow races across Earth's surface.

During totality, the Sun's surface is fully hidden. The main consequence of this is that the human eye can see the much fainter atmosphere of the Sun – the solar corona (see page 25). This beautiful natural phenomenon is the primary reason so many people travel to see total solar eclipses. However, the apparent size of the Moon is not always large enough to fully block out the Sun. But how can the Moon change size?

The Moon is not a constant size in the sky because the shape of its orbit is not perfectly circular, but elliptical. At the farthest part of the Moon's orbit (called apogee), it is 407,000 km (252,898 miles) from Earth. At its nearest point (called perigee), it's much closer at 357,000 km (222,000 miles). This alters the apparent size of the Moon in the sky. The phases of the Moon can happen at any time over the seemingly changing Moon size. If a Full Moon takes place at perigee, we call it a supermoon. If one happens during apogee, it's called a micromoon. A supermoon can have an apparent diameter 14 per cent wider than a micromoon.

> *This beautiful natural phenomenon is the primary reason so many people travel to see total solar eclipses.*

Solar eclipses happen during a New Moon. If the New Moon is closer to perigee (and therefore larger than the average Moon size in the sky), it is large enough to fully block the Sun and cause

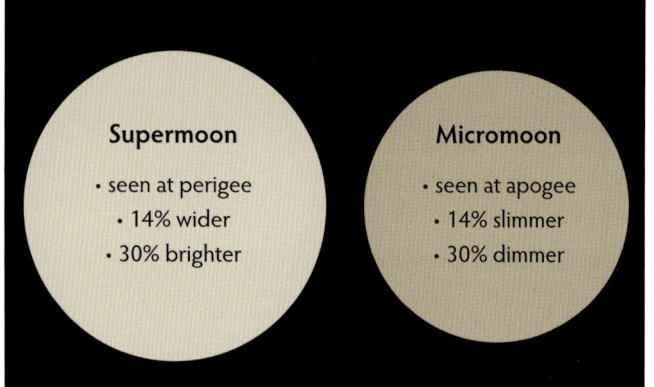

Comparison of the size of a supermoon and a micromoon.

Supermoon
- seen at perigee
- 14% wider
- 30% brighter

Micromoon
- seen at apogee
- 14% slimmer
- 30% dimmer

a total solar eclipse (if the alignment between the Sun, Moon and Earth is right). If, on the other hand, a solar eclipse happens while the New Moon is at apogee, the Moon won't be large enough to fully encompass the Sun in the sky. The result is that, as the Moon passes in front of the Sun, a ring of sunlight fully encircles the Moon. This creates a 'ring of fire' or 'annular' eclipse (see page 102). Just as total solar eclipses are seen for a few minutes within the path of totality, annular eclipses similarly have a path of annularity – the region across which the ring of fire can be seen. As with totality, annularity lasts a few minutes at any location, and it takes several hours for the region of annularity to move along the eclipse path. The diagram (below right) highlights the differences between a total and annular solar eclipse.

Under rare circumstances, the Earth can experience a hybrid solar eclipse. These begin as annular eclipses at the start of the eclipse path, transition to total solar eclipses at the middle of the eclipse path, then revert back to annular eclipses at the end. At any specific location along the eclipse path, an observer will experience either totality or annularity, but not both. Hybrid eclipses happen during the rare occasion that the Moon is at the threshold distance required to transition from one eclipse type to another. Let's imagine the scenario. If you are standing outside at midday, then you are closer to the Sun and New Moon than somebody simultaneously standing in a different time zone during their local sunrise or sunset. This difference in the distance to the Sun and New Moon between these locations will roughly equal the radius of the Earth. This is the basis of a hybrid eclipse. Within regions of the eclipse path closer to sunrise and sunset, where the eclipse is low in the sky, the relative size of the Moon is too small to fully block out the Sun – so they experience an annular solar eclipse. However, within regions of the eclipse path closer to noon (midday local time), the small reduction in distance between the observer and the Moon makes it appear large enough to fully block the Sun and create a total solar eclipse. The Moon's shadow moves much faster than the Earth rotates, so as the hybrid eclipse path races across the Earth's surface, the eclipse switches from annular, to total, to annular solar eclipse.

❝ Under rare circumstances, the Earth can experience a hybrid solar eclipse. ❞

Although the totality and annularity phases of a solar eclipse last only for a few minutes, the entire eclipse lasts for several hours. This is because a solar eclipse does not begin and end with totality or annularity; these occur during the middle of the eclipse. Within the path of totality, you will witness the slow onset of the Moon passing in front of the Sun, blocking out progressively larger and larger areas of the Sun's surface. This is the partial eclipse phase. After totality, this same process happens in reverse. Overall, a total eclipse typically takes a few hours, with the maximum eclipse (totality) lasting only a few minutes in the middle of this duration.

Totality	Annularity

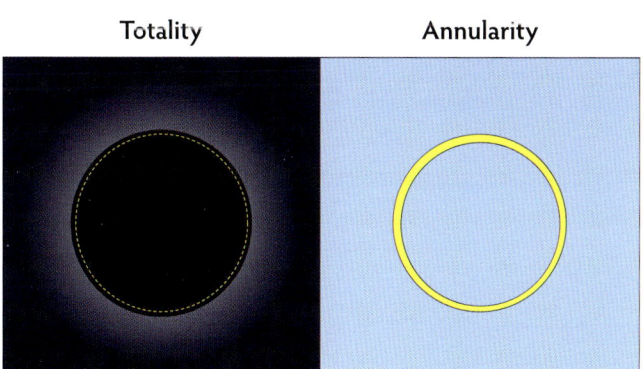

The difference in the size of the Moon during a total solar eclipse and an annular solar eclipse.

Although the path of totality (or annularity) is narrow, just a few hundred kilometres across, the eclipse path can stretch thousands of kilometres across the Earth's surface. If you are outside the path of totality, you may still see a partial solar eclipse – the Sun will be partially, but never fully obscured by the Moon. Partial eclipses are usually represented as percentages. For example, if you are outside of, but close to, the path of totality, you may witness a 95 per cent partial eclipse (meaning 95 per cent of sunlight is blocked by the Moon). This number decreases with distance from the central eclipse path, extending outwards until none of the Sun is blocked from view. This means that even thousands of kilometres away from the path of the total solar eclipse, partial eclipses of 10 per cent can still be visible. Be careful though. The phrase '95 per cent partial eclipse' may sound like you'll have 95 per cent of the experience of a 100 per cent total eclipse – but this is not the case.

> *The comparison between totality and a 95 per cent partial event is literally the difference between night and day.*

The comparison between totality and a 95 per cent partial event is literally the difference between night and day (see page 95). Annular eclipses are a special type of partial eclipse, as not all of the Sun is blocked from view.

Most total and annular eclipses are categorized as central eclipses – an eclipse during which the point of maximum eclipse intersects the Earth's surface. This point can be found by drawing a straight line through the Sun and Moon. If that line intersects with the Earth at any point, it's a central eclipse. Totality and annularity occur at the point of maximum eclipse, but not all total and solar eclipses are central eclipses. The majority are, but there are some exceptions. For some eclipses, the point of maximum eclipse can be just above the top of the Earth (near the North Pole), or just below the bottom of the Earth (near the South Pole), narrowly missing an intersection with the Earth's surface. Although these are not central eclipses (as the point of maximum eclipse is not on Earth), the point of maximum eclipse can be close enough to Earth that regions towards the poles can experience totality or annularity. These rare events are considered to be total or annular solar eclipses, but not central ones.

If the path of maximum eclipse misses the Earth by a sufficient distance, then some regions will observe a (non-annular) partial eclipse, but nowhere on Earth will experience a total or annular solar eclipse. These are 'partial eclipses' and provide the same experience as regions experiencing partial eclipses during a total eclipse, and the partial phases before and after the total eclipse phase along the path of totality. Sole partial eclipses, where nowhere experiences totality or annularity, always occur at high latitudes towards the North or South Pole.

CATEGORIES OF LUNAR ECLIPSES

Lunar eclipses occur as the Full Moon passes into the Earth's shadow. On a typical Full Moon, the Moon is fully basked in sunlight. But, if the alignment of the Sun, Earth and Moon is just right (which only happens during a Full Moon), the Moon can pass fully into the Earth's shadow. The result is a major difference between a solar and lunar eclipse. During a solar eclipse, the observer stands within the shadow of the Moon, looking back towards the Moon and eclipsed Sun. With a lunar eclipse, instead of the observer being within the shadow, the Moon is in the shadow, and we witness this from Earth. A major consequence of this is that whereas the level of solar eclipse you experience depends on where you are in the world, everyone with a view of the Moon will experience the same lunar eclipse – because we are all looking at the same Moon.

Lunar eclipses come in three phases – penumbral, partial and total (see pages 65–76). All total lunar eclipses are preceded by the penumbral and partial phases (in that order), before reversing back through the partial and penumbral phase on the way out of the eclipse. Alongside most total and annular solar eclipses, total lunar eclipses are an example of a central eclipse (see page 54), as the Moon must be within the region of maximum eclipse to produce a total lunar eclipse. If the alignment of the three objects is not close enough to create a total lunar eclipse, partial eclipses (which start and end with the penumbral phase) or penumbral eclipses can occur. Unlike total solar eclipses, which cause totality for just a few minutes at a given location, the total phase of a total lunar eclipse can last for well over an hour, combined with more than an hour of partial and penumbral phases before and after the total phase. This gives observers far longer to enjoy a lunar eclipse. Any location in the world with visibility of the Moon during the extent of the lunar eclipse will witness it. Unfortunately, if the lunar eclipse ends before the Moon rises or after the Moon sets in your location, you will miss it.

So, what happens during a lunar eclipse? To understand this, it is important to remember that the Earth has two types of shadows. Because the Earth has width, it casts both a partial shadow (called the penumbra) and a complete shadow (called the umbra) from incoming sunlight (see the diagram below).

These two distinct shadow regions determine what we see during a lunar eclipse. As the Moon enters the Earth's

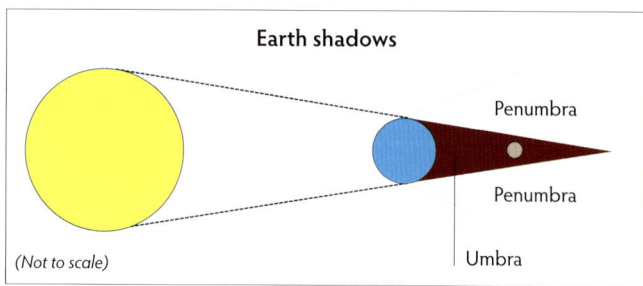

Diagram showing the Earth's umbral and penumbral shadow.

Earth shadows

Penumbra

Penumbra

Umbra

(Not to scale)

three | UNDERSTANDING ECLIPSES

penumbra, we experience a penumbral lunar eclipse, a slight drop in brightness that can be difficult to perceive. As the Moon begins to enter the umbra, the next eclipse stage becomes more noticeable. If part of the Moon sits within the Earth's umbra, we experience a partial lunar eclipse and part of the Moon becomes very dim as the Earth casts its shadow on a section of the Moon. Finally, as the Moon fully enters the Earth's umbra, it creates a total lunar eclipse.

What happens during a total lunar eclipse can be counter-intuitive. During a total solar eclipse, the Sun is fully blocked from view. With a total lunar eclipse, we might expect a similar disappearance of the Moon, but this is not the case. Although part of the Moon appears to vanish during a partial lunar eclipse, it has not disappeared completely, but instead is just far fainter in comparison to the regions of the Moon still outside the Earth's umbral shadow. As more of the Moon enters the umbra, our eyes can adjust to see more of the dimly lit Moon. It appears dark and blood-red, very different to the bright white Full Moons we're used to experiencing. Why is this? The red colour of a total lunar eclipse is caused by sunlight passing through Earth's atmosphere (see page 69).

UNDERSTANDING ORBITS

The final question we must answer here is: Why don't eclipses occur all the time? Eclipses happen due to the fortuitous alignment of the Sun, Earth and Moon. Solar eclipses take place at New Moon, and lunar eclipses at Full Moon. But if the Moon orbits the Earth relative to the Sun once every 29.5 days, why don't we experience one solar eclipse and one lunar eclipse every month? The answer comes down to the precise orbits of the Earth and Moon.

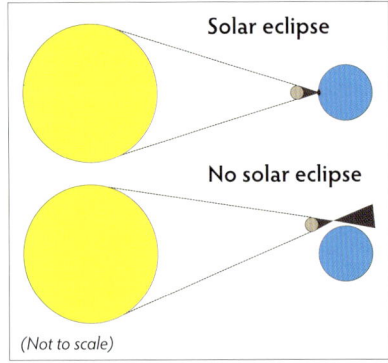

Solar eclipse

No solar eclipse

(Not to scale)

Diagram highlighting the tilt of the Moon's orbit.

The orbit of the Moon around the Earth is not aligned in the same plane as the Earth around the Sun. They are offset by an angle of 5.15 degrees, highlighted in the diagram (below right).

Although this tilt seems small, it is sufficient to ensure that the Moon, Earth and Sun do not typically align at a New or Full Moon, depriving us of monthly solar and lunar eclipses. Instead, opportunities for eclipses only arise twice a year, during the so-called 'eclipse seasons'. These happen when the tilt of the Moon's orbit sits perpendicular to the Sun-Earth line, allowing the Moon to enter the same plane as the Earth and Sun. Each eclipse season lasts around 35 days, and they are the only windows during which solar and lunar eclipses can take place. Eclipse seasons are spaced by just under six months, at an average rate of once every 173 days. This means that we often see some kind of eclipse (solar or lunar, and total or partial) every six months, and successive years may see eclipses in the same month. Because the duration of the two eclipse cycles (173 x 2 = 346) is less than a full year, the eclipse seasons slowly drift through the calendar year as each year passes. The years 2023 and 2024 both saw total solar eclipses in April, whereas 2026 and 2027 will see total solar eclipses in August. Over time, the eclipse seasons shift far enough through the year to end up back where they started. The cycle takes a total of 6,585 days (more than 18 years), and is called the Saros cycle. After each Saros cycle, eclipses return to a similar date and location. Although the cycle also brings a solar eclipse back to roughly the same region, the shadow of the Moon is still narrow enough such that a total solar eclipse happens only once every 375 years (on average) at any specific location on Earth. Typically, a total solar eclipse happens somewhere around the world every 1.5 years.

So, these events are not necessarily rare, but they are scarce at a location near you.

The length of an eclipse season means that we often see back-to-back solar and lunar eclipses (even if only partial) during a given window, separated by two weeks during subsequent New and Full Moons. The diagram (below) demonstrates how the alignment of the Moon's orbit with the Earth and Sun creates these eclipse windows.

The Earth orbits anticlockwise around the Sun, and the Moon anticlockwise around the Earth. Therefore, when watching a solar eclipse from the northern hemisphere, the Moon appears to drift from right to left across the Sun. Because of this, the shadow of the Moon during a total solar eclipse traverses from west to east across the Earth's surface. There is an exception to this. Because the Earth rotates on a tilted axis of 23.44 degrees, a total solar eclipse near the North or South Pole can project the Moon's shadow over the poles to the other side of the Earth. On this 'backside' of the Earth, the Moon's shadow appears to move from east to west.

The west-east direction of the path of totality is determined by the Moon's motion, but what determines whether the Moon's shadow approaches from the north-west or south-west? This comes down to eclipse seasons again. During one eclipse

Diagram demonstrating eclipse windows throughout Earth's orbit around the Sun.

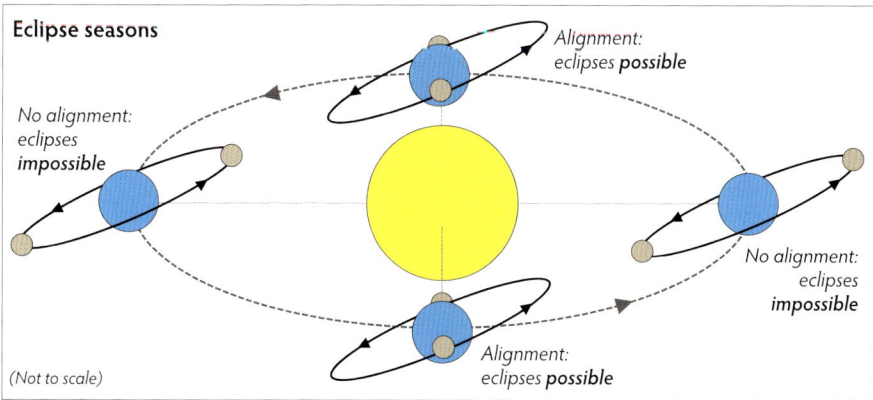

Eclipse seasons

No alignment:
eclipses
impossible

Alignment:
eclipses **possible**

No alignment:
eclipses
impossible

Alignment:
eclipses **possible**

(Not to scale)

season, the path of the Moon moves upwards into the Sun-Earth line at New Moon. During the alternative eclipse season (six months before and after) the Moon passes downwards into the Sun-Earth line. As a result, eclipses six months apart will have an eclipse path travelling from opposite north or south directions.

During a total solar eclipse, the shadow of the Moon pans faster than the Earth can rotate. If the Earth's rotation were much faster, then eclipse paths would instead stretch from east to west – but this is not the case. The speed of the Moon's shadow depends on where on Earth the eclipse is happening. The shadow of the Moon moves the slowest around local noon at the equator, travelling at speeds of 1,600 km per hour (1,000 miles per hour) or so. Closer to the poles, or over the 'sides' of the Earth close to sunrise or sunset, the projection of the Moon's shadow means that it moves much faster – upwards of 8,000 km per hour (5,000 miles per hour).

An observer at a location where the Moon's shadow travels more slowly will be in the shadow of the Moon for longer and therefore experience totality for more time than someone under a faster-moving Moon shadow. This speed of the Moon's shadow, combined with the area of the Moon's shadow (which is determined by the apparent size of the Moon in the sky) governs how long totality persists for. The longest the totality period of a total solar eclipse can last is a little over 7 minutes, but these are incredibly rare. The longest eclipse of the 21st century was an eclipse in 2009, with a totality of 6 minutes and 39 seconds. The total solar eclipse in 2027 (one Saros cycle after the 2009 eclipse) will have a totality of 6 minutes and 23 seconds. This is still a long solar eclipse, as many have durations of just 1 to 4 minutes.

Diagram showing the direction of the Moon's shadow during a total solar eclipse.

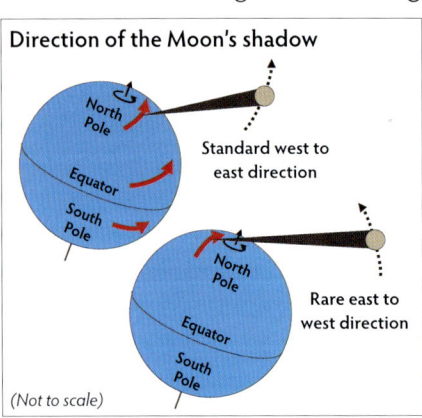

Direction of the Moon's shadow

North Pole

Standard west to east direction

Equator

South Pole

North Pole

Rare east to west direction

Equator

South Pole

(Not to scale)

four

LUNAR ECLIPSES

O F THE TWO FAMILIES of eclipses, lunar eclipses are by far the more accessible. As you'll see in the wide coverage of the eclipse maps in chapter 7 (see pages 123–152), potentially billions of people have the opportunity to enjoy a typical lunar eclipse. For any location, they are mercifully frequent. You can expect a deep partial or total lunar eclipse where you live every couple of years. By comparison, central solar eclipses are statistically unlikely to visit you, gracing a given location on average once every few centuries. These solar events are also fleeting, offering just minutes of totality or annularity, whereas total lunar eclipses can last as long as 1 hour and 47 minutes – a generous period in changeable weather.

> *You can expect a deep partial or total lunar eclipse where you live every couple of years.*

Lunar eclipses are visible over large swathes of the Earth's surface and are perfectly safe to observe without any special equipment.

→

OBSERVING LUNAR ECLIPSES

Fortunately, lunar eclipses can be enjoyed without any special equipment. A lunar eclipse occurs during a Full Moon, which can be readily appreciated with the unaided eye. At total brightness, a Full Moon is also safe to observe directly with binoculars or a telescope, so all forms of lunar eclipse can be enhanced using an optical aid. Indeed, even a small pair of binoculars will significantly brighten the appearance of the Moon to your eyes, which can offer a counterbalance to the natural darkening brought about by the Earth's shadow.

The most significant challenge, aside from weather, is the subtlety displayed by penumbral or shallow partial eclipses. In such cases, the Earth's shadow can be hard to detect at all.

During a penumbral eclipse, evidence of the Earth's shadow is scarcely visible unless you know what to look out for. In this case, it is darkening the southern reaches of the Moon.

It's hardly surprising then, that these events often go entirely unnoticed by the casual onlooker. They certainly can't be considered spectacular, but they are nevertheless worth looking out for – if only for the rewarding challenge of spotting the shadow.

As, even at full brightness, it is safe to look at the Full Moon with binoculars or a telescope, you may find yourself admiring the face of our celestial neighbour during an eclipse. If so, plenty of interesting sights come into view under magnification, and there are several curiosities to look out for regardless of how you choose to view the event.

Though the Moon has a small angular size of about 0.6 degrees, deep eclipses are very arresting to the unaided eye.

Binoculars and small telescopes reveal most of the Moon's major features and craters, labelled on this map. How many can you find?

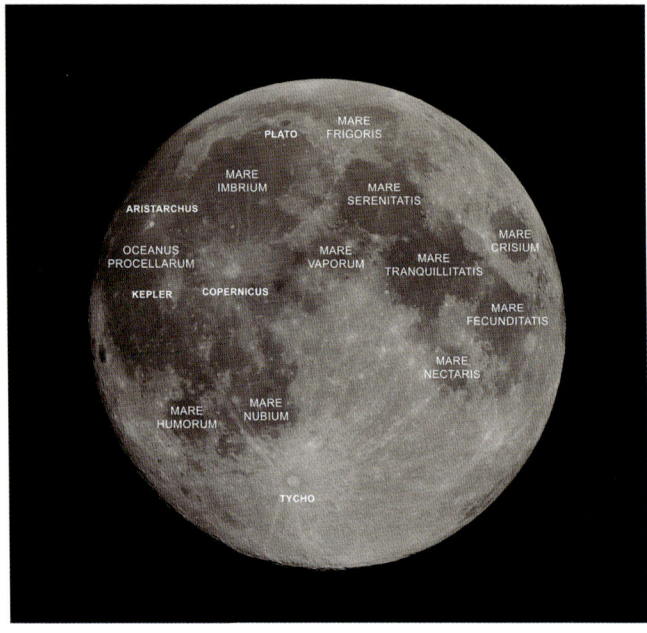

UMBRAL HUES

The unique element, which gives any good partial or total lunar eclipse its unmistakable character, offers one of the most compelling opportunities in astronomy. It is a glimpse of the Earth's atmospheric optics on a grand scale. If the Full Moon passing through the centre of the Earth's shadow was all you knew about total lunar eclipses, in the absence of any pictures you'd be forgiven for assuming that the Moon's face would become very dark, perhaps even invisible to the eye. However, the umbral region of the shadow is not pitch black. It is flooded with sunlight that is filtered through the densest regions of the atmosphere.

As a sunset washes the sky in warm tones, deepening as darkness approaches, so the Moon also becomes increasingly red the closer it comes to the centre of the umbra. There is some poetry to the comparison, since the eclipsed Moon is being

A gradient of red and orange light, darker and brighter respectively, transforms the grey canvas of the lunar surface during a total lunar eclipse.

↑

This high dynamic range photo composite reveals the dramatic difference in brightness between the penumbral and umbral shadows. For this reason, a lunar eclipse typically goes unnoticed by many onlookers until the partial phase begins.

painted with the light of all the world's sunrises and sunsets. The resulting spectrum of hues across this rocky canvas may include ruddy scarlet, coppery brown or pale amber, and everything in between.

To understand these tints, we must first recall that sunlight is broad-spectrum white light, composed of photons of every visible colour. When this light interacts with atoms or molecules in the atmosphere – principally nitrogen and oxygen – it is subjected to a variety of physical processes. The atmosphere effectively filters and redirects sunlight, sending a selective portion of it into the umbral shadow.

The filtering occurs due to scattering. In this process, photons interact with atoms, molecules or very small particulates, causing them to be absorbed and re-emitted. Since the emission direction is effectively random, light is redistributed by this process. Scattering is a wavelength-dependent phenomenon, and physicists have classified its effects:

→ **Rayleigh scattering** occurs when a photon encounters a particle somewhat smaller than its own wavelength. The atoms and molecules in the atmosphere have sizes and spacings smaller than blue photons (the shortest wavelength visible light), so Rayleigh scattering casts blue light all over the sky.

→ **Mie scattering** occurs when a photon encounters a particle comparable to or larger than its own wavelength. Some particulates in the atmosphere have sizes and spacings similar to red photons (which have longer wavelengths), and Mie scattering causes some deviation in the direction of red light. However, the effect is considerably less pronounced than Rayleigh scattering.

Since Rayleigh scattering randomizes the direction of blue light, as sunlight passes through the atmosphere its blue component is filtered. When the Sun is low on your horizon, from your point of view its light must pass through considerably more atmosphere than when it is overhead. By the time it reaches you, so much of its blue component has been filtered that it is left with a strong red bias. Mie scattering subtly alters the direction of red sunlight, which spreads out through the atmosphere. As a result, a fraction of the red light finds its way around the curvature of the Earth and into the shadow. From a sufficient distance on the nightside, as the Earth eclipses the Sun, you can see a red ring around it. When the Moon is submerged in the Earth's umbral shadow, this red ring is the brightest light source in the sky.

The umbral shadow's colour and darkness are not fixed, and no two umbral lunar eclipses are exactly alike in either aspect. This is because the scattering profile of the atmosphere changes with its composition. When the air is relatively clear, globally speaking, the Moon is a striking bright orange, or red-orange, as it passes through the umbra. Major events, such as volcanic eruptions, large-scale wildfires or seasonal dust storms, introduce particles into the atmosphere, increasing the Mie scattering factor. When this happens, the eclipse will generally be darker and less vibrant, taking on a dull or muddy colour. Just as Rayleigh scattering reddens the appearance of the Sun when it is low on the horizon, the rising or setting Moon can also appear warm-toned.

As sunlight is filtered through the atmosphere during an umbral eclipse, its hue shifts. The result is a range of warm tones that paint the lunar surface.

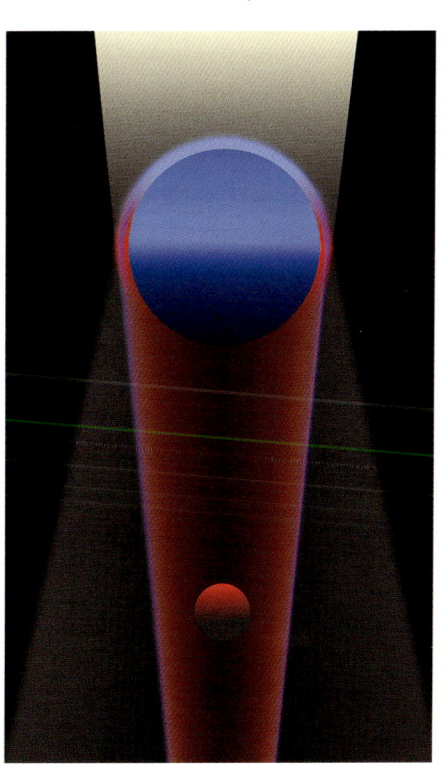

→

In this composite image, captured shortly after the end of the total phase, the brightness has been normalized and the colours enhanced to reveal the extraordinary range of hues passing through the atmosphere. The violet ozone absorption band can be seen on the left.

Some sky-watchers mistake an ordinary Full Moon for an eclipse when it is skirting the horizon. Consequently, an umbral eclipse will appear particularly red and dull if it occurs around moonrise or moonset from your location, as sunlight is filtered once on its way to the Moon and again on its way back to you.

You might be fortunate enough to see an entirely different colour. It's a real treat, but it typically appears subtle and therefore requires a sharp eye to make out, which is why it often goes unnoticed. At the boundary of the umbra and penumbra, a greyish-white fringe can appear. Look closely and you'll see it's violet or turquoise. The red and orange tones are produced by the lower and upper regions of the troposphere, respectively. Above that, around 10 km (6 miles) above the ground, is the stratosphere, which begins with the ozone layer. Ozone is well-known for its absorption of ultraviolet light. However, in the visible part of the spectrum, its absorption profile skews towards red, and blue light is transmitted through it. As a result, the ozone layer seems to provide an almost inverted filter to the troposphere.

THE VIEW FROM
THE MOON

Once you have seen the colour of the umbral shadow, it is possible to picture how the Earth might appear to an observer standing on the Moon's near side during a total lunar eclipse. Or, to put it another way, what a total solar eclipse looks like from the Moon. For a number of reasons – most notably the huge variation in surface temperature – this was not considered to be a suitable experience for any of the Apollo astronauts (see page 28). From the perspective of someone standing on the Moon, and based on the geometry of the eclipse, the Sun would be entirely obscured, and the Earth's atmosphere would appear as a bright, red-tinged ring in the otherwise starry sky. In 2025 the imaginations of astronomers were finally vindicated when Firefly Aerospace successfully achieved the first commercial soft landing on the lunar surface. The company's Blue Ghost Mission 1 deployed its lander in Mare Crisium on 2 March 2025, and on 14 March, a total lunar eclipse swept across the Moon.

Left: NASA's Surveyor 3 captured the atmospheric ring during the lunar eclipse on 24 April 1967.

Right: Apollo 12 astronauts (two of whom had visited Surveyor 3 on the surface) passed through the Earth's shadow on their way home. This photo was taken around 23 November 1969.

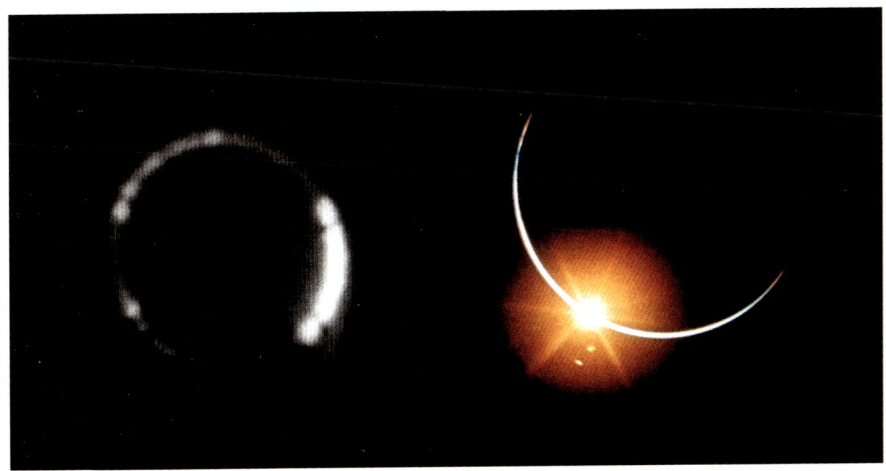

↑

Left: **The Firefly Aerospace Blue Ghost lander** returned this photo of a spectacular diamond ring (see page 94), with the atmospheric ring revealed in colour.

Right: Red umbral light paints the Blue Ghost lander. Brilliant Venus is seen above the eclipse, with fainter Mercury just to the left of it.

While it wasn't the first solar eclipse captured from the Moon (that honour goes to Surveyor 3 in 1967) or the first colour photo from within the Earth's far shadow (taken by astronauts aboard Apollo 12 in 1969) the images represent the first time the Earth's atmospheric ring was captured in colour from the lunar surface. What's more, the lander also observed itself bathed in the red light we see reflected back to us on Earth.

> ' *While it wasn't the first solar eclipse captured from the Moon . . . the images represent the first time the Earth's atmospheric ring was captured in colour from the lunar surface.* '

IMPACT EVENTS

Every so often, the surface of the Moon is struck by impactors – various rocks that make up the sparse debris in interplanetary space – which result in bright flashes and the formation of new craters. The day-lit side of the Moon is so bright that the flashes can't be seen, even with a telescope. However, during a lunar eclipse, the shaded face of the Moon is dark enough to show much greater contrast with bright sources, like those caused by impact events. Since eclipses have become so widely and attentively watched around the world, it was inevitable that the formation of a new crater would be captured in real-time. That occurred during the total lunar eclipse of 21 January 2019. At 4:41:43 UT, as the total phase began, multiple observatories around the world observed a flash near the Moon's western limb. A camera at the Royal Observatory in London, England, also detected a second flash near the eastern limb 2 minutes later at 4:43:44 UT. As a result of thin cloud cover further reducing the brilliance of the lunar surface, it is the only instrument to have done so.

In January 2019, a 45 kg (99 lb) rock struck the Moon at approximately 61,000 kph (38,000 mph) releasing energy equivalent to 1.5 tonnes (1.6 tons) of TNT, enough to create a crater roughly 15 m (49 ft) wide. The impact (*left*) occurred on the western limb. Another impact (*right*) caused a bright flash on the eastern limb 2 minutes later.

↓

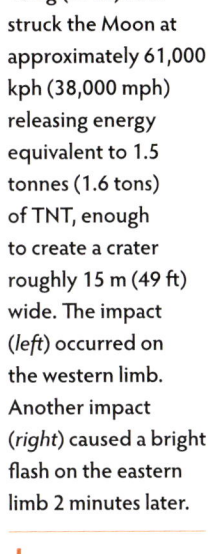

This first sighting of a pair of impacts on the Moon opens up a new observational opportunity, albeit a very challenging and extremely rare one. If you keep your eye to the telescope, or better yet let a camera monitor in your stead, you may one day join a short and exclusive list of eclipse-chasers who have witnessed the transformation of the Moon's surface as it happens.

STARGAZING

Ordinarily, the Full Moon represents a significant source of natural light pollution. It brightens the sky around it with a halo of moonglow scattering through the Earth's atmosphere, reducing the surrounding contrast and washing out fainter stars and other objects. For this reason, many professional observatories avoid observing parts of the sky anywhere near the Moon. However, the drastically reduced brilliance that results from a deep lunar eclipse makes stargazing in the vicinity of the Moon possible again. You can readily photograph it with the surrounding stars preserved. Comparing the average brightness of uneclipsed moonglow to the sky around a total eclipse Moon yields a difference of about three stellar magnitudes.

There are numerous interesting stars and constellations flanking the ecliptic path, which the Moon is necessarily always close to. There are also notable star clusters, such as the Pleiades and Hyades in Taurus, the Bull, and the Beehive Cluster in Cancer, the Crab. Ordinarily, they would become lost in bright moonglow, but they can add to the celestial magic of a lunar eclipse.

The lunar eclipses detailed in this book (see pages 127–135) are accompanied by charts that illustrate the locations of constellations around the Moon, as they would appear during greatest eclipse from the sublunar point. The view will vary from other locations and at other times, but they will nevertheless help you identify the night sky in the vicinity of the Moon.

To be visible, faint stars must show a good amount of contrast with the background sky. During a total lunar eclipse, the halo surrounding the Moon virtually disappears, allowing stars to be seen and photographed close to the lunar limb.

A TOTAL SOLAR ECLIPSE is perhaps the most spectacular event visible to us from Earth. Although rarer and less widely accessible than lunar eclipses, they are orders of magnitude more impressive to witness. A total solar eclipse happens somewhere around the world every one and a half years, but only at any specific location (on average) once every 375 years. This means that the majority of the world's population has to travel to see a total solar eclipse. The trip may be long, but it's certainly worth doing at least once in a lifetime.

OBSERVING SAFELY

You should never look directly at the Sun. Doing so can cause irreversible harm to your eyes. This can manifest itself in several different ways, including by damaging the parts of your eye responsible for seeing colour (the cones), permanently desaturating your colour perception. Staring at the Sun could also leave permanent blind spots in your vision. You are probably familiar with these blind spots, present when blinking after staring at a bright lamp. Those will vanish after a minute or two, but blind spots from staring at the Sun can last forever. In North America, more than 100 people sought hospital treatment for eye injuries following the 2017 and 2024 total solar eclipses.

The surface of the Sun, called the photosphere (the layer of the Sun we see every day) is bright. Even during a partial eclipse, when a significant portion of the Sun's surface is blocked by the Moon, it is still too bright if viewed directly. This means that for any partial solar eclipse (including annular eclipses), or for the partial phases of a total solar eclipse, you should never look at the Sun directly. This is equally as true

> ❛ A total solar eclipse is perhaps the most spectacular event visible to us from Earth. ❜

People take precautions by using solar filters to view a solar eclipse in Düsseldorf, Germany, in 1914.

for a 99 per cent partial eclipse, as it is for the uneclipsed Sun. Looking at it could cause serious harm, and you won't be able to see anything interesting anyway – the Sun is just too bright. There are many fun and creative ways to safely view and enjoy partial eclipses.

OBSERVING PARTIAL ECLIPSES

In this context, a partial solar eclipse is any solar eclipse where some of the Sun's surface is still visible. This includes standard partial solar eclipses, annular solar eclipses (see page 56), and the partial phases before and after the totality stage of a total solar eclipse. These methods are usable for any percentage of partial solar eclipse (ranging from 1 to 99 per cent of the Sun's surface being blocked by the Moon).

SOLAR VIEWING GLASSES

Probably the most well-known method to watch a partial eclipse, solar viewing glasses (or solar eclipse glasses) block out more than 99.9999 per cent of incoming light, making everything around you (except the Sun) appear jet black. (Regular sunglasses are not a suitable alternative – they only stop around 20 per cent of sunlight.) Eclipse glasses not only block out enough sunlight to make the Sun comfortable to look at, but they also obstruct harmful ultraviolet and infrared light that, although invisible to the eye, can still cause discomfort or damage. Solar eclipse glasses should meet the international safety rating ISO 12312-2, which should be marked somewhere on the glasses when purchased.

Eclipse glasses typically have paper frames (although more expensive versions may use sturdier materials), with lenses made from filters of either a black polymer or silvery mylar material. Both types of lenses are safe, but each one gives the Sun a different colour. With polymer filters, the Sun appears to be orange, whereas mylar gives the Sun a blueish-white tint. These solar viewing filters are used in other devices – handheld solar filters are also common – and it is also safe to use larger sheets of the filter to create attachments for your camera or

NASA astronauts Stephen Bowen, Frank Rubio and Warren Hoburg, and United Arab Emirates astronaut Sultan Alneyadi remind the public to wear appropriate safety glasses to view a solar eclipse by wearing solar viewing glasses before their 2024 trip to the International Space Station.

telescope. Solar eclipse glasses are relatively inexpensive, but may be in short supply in the days leading up to an eclipse. It's best to be prepared and organize your glasses a few weeks ahead of time. In North America, during the 2017 total solar eclipse, eclipse glasses were in such short supply just before the eclipse that dangerous fake examples flooded the market. This caused major health concerns.

In the past, people have asked whether a welding mask, or welding goggles, offer a suitable alternative to eclipse glasses. The American Astronomical Society states that a welding filter of grade 12 or higher does provide suitable protection for looking at the Sun. This is higher than the standard filter grade that often comes with a welding mask. Having tried this out personally, I still found a grade 12 filter to bring discomfort when looking at the Sun.

PROJECTIONS AND REFLECTIONS

If you don't have access to solar eclipse glasses, or are looking for more methods to enjoy the partial phases of a solar eclipse, you can use projections and reflections to view the event. There are multiple ways to do this, and none of them involve looking directly at the Sun. The best thing about these methods is that they are either cheap or free, and most of them use items it is likely you already have, or that are easy to obtain.

The simplest way is to create a pinhole camera. You can make one by poking a small hole through a piece of paper or cardboard. Hold it a few tens of centimetres away from a second sheet of paper or cardboard and a reversed image of the partial eclipse will be projected onto the second sheet. You'll have to experiment with the exact distance at which to hold the two sheets, but the crescent Sun will eventually be visible. The image will be small, but you'll immediately recognize the current phase of the partial solar eclipse.

↑

A grade-12 welding mask *(top)*, and the view of the Sun through it *(bottom)*.

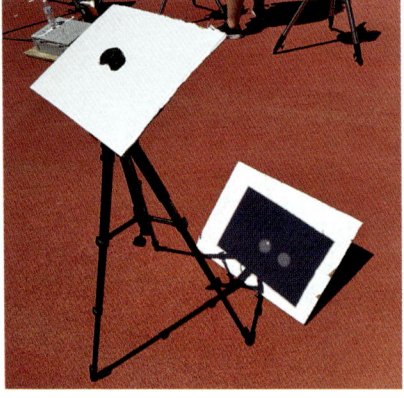

This technique also works with larger holes. This is a fun way to view the eclipse, especially for children, as you can create your own tools to project the eclipse. Whether you cut out a triangular, square or circular hole, the sunlight passing through it will be the same shape as the current stage of the partial solar eclipse. If you hold a colander up to the partially eclipsed Sun, the beams of sunlight passing through its holes will bear the shape of the eclipse when it reaches the ground. The same is true for many objects. If you look under a tree during a partial eclipse, the rays of light that hit the ground through the narrow gaps in the leaves will produce hundreds of crescent-shaped points on the ground. If you have long hair, you can also hold up your hair to create the same eclipse-shaped patterns on the ground.

↑

Top: A partial solar eclipse viewed through a colander.

Bottom: A DIY eclipse projection, made from binoculars, a tripod and some cardboard.

If you want to be more ambitious, you can try this on a larger scale. If you have a regular pair of binoculars, they are **not safe** to use to look at the Sun. However, you can use them to create a safe projection of the eclipse. All you need are two sheets of cardboard, some binoculars and a tripod. Cut a hole in the first sheet of cardboard to match the shape of the front (the side you don't look through) of the binoculars. Attach the binoculars to the tripod (it does not have to be a technical attachment; some tape is fine), and push the binoculars into the cut-out shape in the cardboard to create a frame. At this stage, it is useful to add a sign, telling passers-by **not** to look through the binoculars. Angle the binoculars to point at the Sun but do not look at the Sun through them. Place the second piece of cardboard in the shadow cast by the piece of card attached

to the binoculars, and prop it up at an angle. The two pieces of cardboard need to be close to parallel and about 1 m (3 ft) apart. Take time to fine-tune the angle and distances until you have a projection of the solar eclipse on the second sheet of cardboard. (It will be a few tens of centimetres across.) The projection will be big enough to show the structure of the edge of the Moon and any sunspots (dark spots on the Sun's surface).

Using a disco ball to view a partial solar eclipse (A G M Pietrow, as featured in 'Why every observatory needs a disco ball').

Reflections of an eclipse work in a similar way to projections. A fun way to try this technique to view a partial solar eclipse is to use a disco ball. When the sunlight hits the ball, the hundreds of reflected points of sunlight from it will not be the same round shape we are accustomed to seeing. Instead, every point of light will have the same form as the current phase of the partial solar eclipse. This method works best when the light from the disco ball is reflected into some shade.

SOLAR TELESCOPES AND BINOCULARS

Regular binoculars and nighttime astronomy telescopes are not safe for viewing the Sun directly. They are far more likely to cause permanent damage to your eyes than staring at the Sun with the naked eye. The best way to view a partial solar eclipse is by using a solar telescope or solar binoculars, but these are a more expensive method. They use solar filters, similar to those used in eclipse glasses, to reduce 99.9999 per cent of incoming sunlight to safe viewing levels. These filters, combined with the magnification provided by the telescope or binoculars, enable you to see far greater details on the surface of the Sun, or edge of the Moon, than you can make out with regular (unmagnified) eclipse glasses.

'White-light' solar telescopes are the least expensive and most common models. Much like eclipse glasses, they filter out the majority of incoming light and provide a magnified view of the solar photosphere and a clear image of sunspots and other details. They can be a dedicated solar telescope (manufactured for solar viewing only), or converted from a 'night-time' astronomy telescope by attaching a solar filter to the end of the telescope. If you already have a night-time telescope, purchasing one of these is far cheaper than buying a dedicated solar telescope.

For an even better view of the Sun during a partial solar eclipse, H-alpha telescopes are the way to go. 'H-alpha' stands for 'hydrogen alpha', a narrow wavelength of light produced by a specific reaction of hydrogen plasma in a layer of the lower solar atmosphere called the chromosphere. H-alpha filters filter incoming sunlight, not only by uniformly blocking 99.9999 per cent of light, but by allowing only a thin wavelength range of light (emitted by this chromosphere layer on the Sun) to reach your eye. The result? A beautiful and magnified view of the chromosphere, revealing all sorts of details and structures not visible when using a white-light solar

The view of the Sun through a detachable white-light solar telescope filter.

↓

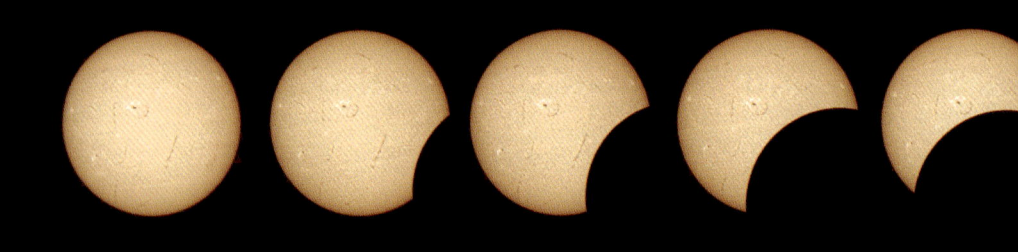

telescope. During the partial eclipse phase, H-alpha telescopes may reveal arches of plasma, called prominences (see page 98), above the Sun's surface. These are visible to the naked eye during a total solar eclipse.

OBSERVING TOTALITY

The solar corona is about 10,000 times brighter than the Sun's surface – a similar brightness to the Full Moon. This difference explains why we can't usually see the solar corona – it's just too faint. Even for a 99 per cent partial eclipse, the remaining thin slice of the Sun is still 100 times brighter than the corona. Unlike the solar photosphere, the corona is completely safe to look at directly, without solar viewing glasses or filters. This means that for the totality phase of a total solar eclipse (and this phase only), you should watch the corona directly with your naked eye. It is important to note that the totality phase lasts only a few tens of seconds to a few minutes, and is both preceded and succeeded by much longer partial eclipse phases. Only when darkness has arrived, and you see nothing at all through your solar viewing eclipse glasses, is it safe to take them off and witness the magic of the solar corona directly. During totality, it is even safe to use regular binoculars to examine the corona in extra detail. When the totality phase ends, and the Sun starts to reappear back behind the Moon – it's time to put your eye protection on again.

A partial solar eclipse viewed from an H-alpha telescope.

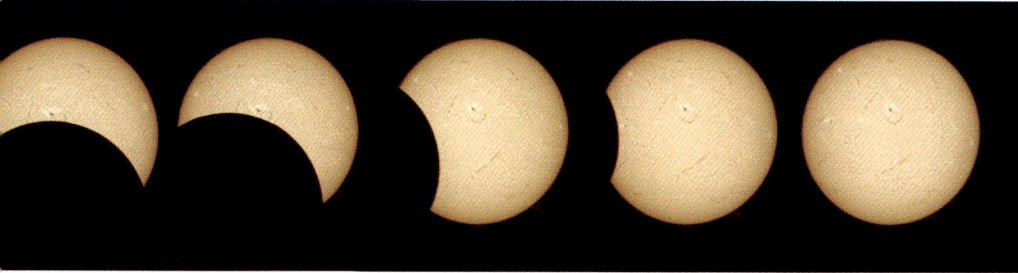

↑

A sketch of the 1806
total solar eclipse
by José Joaquín de
Ferrer.

Now that we know how to safely view the partial and total phases of a solar eclipse, what exactly should we be looking out for?

THE HYPNOTIC SOLAR CORONA

In the hours before totality, during the partial phases of the eclipse, the Moon slowly creeps across the Sun, gradually blocking out the sunlight. As the final slithers of sunlight peak out from behind the edge of the Moon, everything is about to change. The landscape blackens, wildlife quietens and an other-worldly sight can be seen in the sky. An ethereal glow surrounds the jet-black Moon, which appears almost like a black hole punctured through the atmosphere. The light emanates from behind the Moon, filling a large area of the sky. This is what Spanish astronomer José Joaquín de Ferrer (1763–1818) travelled to the USA to see in 1806 (see page 25) and why he named it 'corona' after the Spanish word for crown.

Ferrer was one of the first astronomers to suggest that the corona belonged to the Sun, and not the Moon. He was correct, and the word corona is still used to describe the atmosphere of the Sun to this day.

Although our eyes cannot distinguish it on a sunny day, the Sun's corona is a huge, dynamic structure, several times wider than the Sun itself. The corona is only ever visible to humans during the totality phase of a total solar eclipse, where the Moon blocks the Sun's surface to provide a view of this ethereal structure in the sky above us. This is what Ferrer, and many astronomers before him, voyaged across the world for – and what millions of us travel great distances to see in modern times.

But don't be mistaken. The chance to see the solar corona with your bare eyes, amid the other sensory phenomena provided by a total solar eclipse, is not just for astronomers – it's for everyone. Even if you don't care about astronomy in the slightest, you should still aim to witness at least one total solar eclipse in your lifetime. In the moments of totality, as the Sun's atmosphere sits exposed for us to witness, it's not about astronomy or science. This is because, on a fundamental human level, total solar eclipses are not about the science of the Sun, Moon, or the relative celestial dance between them. Total solar eclipses are about one thing – awe. The sheer raw beauty you experience is unlike anything else. Partial solar eclipses are amazing, but a total solar eclipse will stay with you for life.

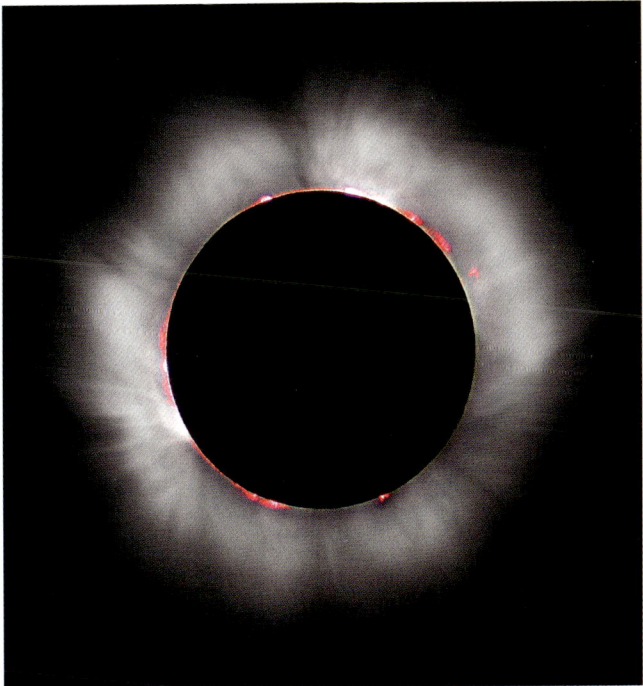

The solar corona visible during the 1999 total solar eclipse.

ECLIPSE CHECKLIST

Although the solar corona is the main marvel to see during a total solar eclipse, it is far from the only phenomena worth paying attention to. Here is an itemized list of all the phenomena visible during a solar eclipse. Some of them are visible during the partial eclipse phases, but the most impressive can only be seen during totality.

	Partial solar eclipse	Strong partial solar eclipse (85 per cent+)	Total solar eclipse	Annular solar eclipse
First/last contact	■	■	■	■
Crescent Sun	■	■	■	■
Sunspots (if present)	■	■	■	■
Crescent shadows	■	■	■	■
Getting darker		■	■	■
Temperature drop		■	■	■
Animal behaviour		■	■	■
Sharper shadow		■	■	■
Colour changes		■	■	■
Shadow bands			■	
Incoming shadow			■	
Baily's beads			■	
Diamond ring			■	
Racing shadow			■	
Corona			■	
Corona structure			■	
Filaments and chromosphere			■	
Dark Moon			■	
Sky colour			■	
Planets and stars			■	
Your friends			■	
Devil horns				■
Annularity				■

EARLY PARTIAL PHASE

A solar eclipse begins at first contact, the moment the edge of the Moon 'touches' the edge of the Sun for the first time. Of course, the Moon doesn't really touch the Sun, but it certainly appears that way from our vantage point on the ground. At this stage, there isn't anything to see with the naked eye. However, with a pair of eclipse glasses (or any other eclipse-viewing method, see page 79), you'll notice the eclipse has begun. Over the next hour or so, the Moon slowly blocks a progressively larger area of the Sun. You cannot see the Moon directly, but you can see the Sun's surface with a Moon-shaped chunk taken out of it to create a crescent Sun.

With enough magnification (such as from a solar telescope), the edge of the Moon will not look like a perfectly smooth circle. Instead, you might see its jagged edge against the bright Sun, formed by the thousands of craters along the visible edge of the Moon. These will become more important later in the moments before totality (see page 92). If they're present, and you have the equipment (see page 83) to see them, sunspots on the Sun's surface, or filaments suspended in the Sun's atmosphere, are also visible during this stage.

Minutes after first contact in a solar eclipse (*left*). A partial solar eclipse, revealing details along the Moon's edge. Sunspots can also be seen (*right*).

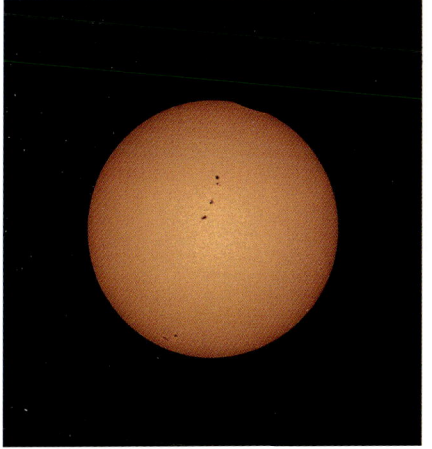

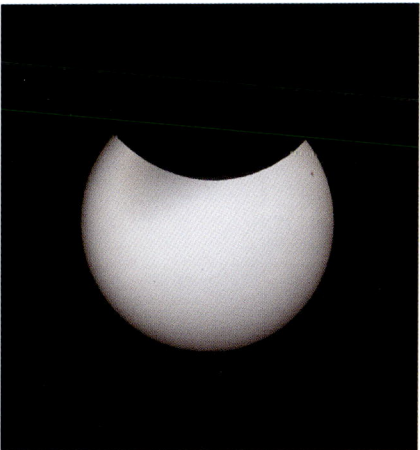

Let's take our attention away from the Sun, and focus on the world around us. For most of the partial eclipse phase – until around 85 per cent of the Sun is blocked by the Moon – everything appears unchanged to the naked eye. This seems counterintuitive, as during a 50 per cent partial eclipse, sunlight levels are half what they are typically. So why does nothing look any different? The human eye is fantastic at adjusting to lower light levels. As the Moon slowly blocks out more of the Sun, our pupils expand in reaction to the lowering light levels faster than the Moon travels across the Sun. This means that although the world around us is darkening, we don't notice it. This all changes beyond an 85 per cent partial eclipse.

DEEP PARTIAL PHASE

When a partial eclipse goes beyond 85 per cent – its deep phase – you'll begin to notice changes in the environment around you. Temperatures typically drop by a few degrees, and light levels become noticeably darker (as your pupils are almost fully dilated). If you are surrounded by nature, wildlife in the area will mistake the changing conditions for twilight, and may

Crescent-shaped shadows cast by a tree.

exhibit behaviour unusual for the time of day. Daytime birds will stop singing, bats and owls may emerge, insects vanish, and so on. Domestic chickens and wild fowl might go home to roost, and even cows and horses begin their nighttime routine. This behaviour will all accelerate towards totality. If you're in an urban area, automatic street and building lights may turn on.

> *If you are surrounded by nature, wildlife in the area will mistake the changing conditions for twilight, and may exhibit behaviour unusual for the time of day.*

During the strong partial phase of a solar eclipse, or perhaps earlier if you're observant, you'll see crescent-shaped shadows around you. Shadows cast by leaves on a tree (or other objects with narrow gaps) will bear the shape of the current eclipse phase (below left). These crescent shadows are worth looking for, as they are easy to miss during the excitement of an eclipse.

As the eclipse inches closer to totality, you'll start to experience a few more strange phenomena that are only present during eclipses. On a sunny day without an eclipse, the Sun has 'width' in the sky. That is, sunlight does not come from a single 'point', but a slightly wider area. As a consequence, shadows cast by sunlight are not perfectly sharp; their edges are soft. In contrast, during some strong partial eclipses, a thin slither of the Sun casts sharp shadows. This effect is difficult to capture in photographs, but it is something our brain immediately recognizes as unusual. But not every strong partial eclipse creates sharper shadows. Some eclipses, such as annular solar eclipses (see page 102), create a long slither around the edge of the Sun (instead of the short one needed to cause sharper shadows), so has enough 'width' in the sky to cast more regular shadows.

In addition to the unusual shadows, your brain will notice an uncanny colour change. This happens during very strong partial eclipses, and in the minutes preceding (and succeeding) totality. Some eclipse observers describe the colour shift as grey or silver. Regardless, it will look as if a colour filter has been applied to the world around you. This is an effect of the human eye. As light levels lower, the eyes' sensitivity to colour shifts from longer wavelengths (the red end of the visible spectrum) to shorter wavelengths (the blue end). This is called the Purkinje effect. This is one of the less frequently discussed phenomena during eclipses (perhaps because it is less exciting than the other things to watch out for on this list), but it is definitely worth noting.

MOMENTS BEFORE TOTALITY

We are now a minute or two away from totality, with only a tiny slice of the Sun still visible from behind the Moon. At this stage, when the remaining sunlight comes from a tiny area in the sky, something strange appears on the ground – shadow bands. These are thin (a few centimetres across) streaks of shadows that move quickly across the ground. Their exact cause was uncertain for a long time, but they are now believed to be due to turbulence and gravity waves in the Earth's atmosphere (see page 46). In the night sky, stars twinkle due to the turbulent air wobbling as the light from a star passes through it. When the perfect fraction of the Sun is eclipsed before and after a total solar eclipse, this turbulent air causes the Sun to twinkle. Although the direct glimmer of the Sun cannot be perceived by our eyes, the subsequent shadows cast by the twinkling eclipse are. Shadow bands are very difficult to view in ordinary environments, but can be seen easily on blank white surfaces. If you want to catch a glimpse of these elusive shadow bands, place a large white sheet on the ground, or hang it up where the Sun will hit it and wait for them to appear.

Shadow bands visible moments before totality.

Totality is now 30 seconds away. Because the shadow of the Moon is moving so fast (2,414 km per hour / 1,500 miles per hour for an average eclipse), the next few items on the checklist happen quickly. The Moon's shadow (and region of totality) travels from west to east for most eclipses. At this speed, if totality is only 30 seconds away, the Moon's shadow is just 19 km (12 miles) away. Looking up at the sky, you can see totality coming. To the east, the sky is blue and bright. But to the west it is dark as you look towards the shadow of the Moon. This is the incoming shadow, and you can see it racing towards you from the west, moments before totality. If you are viewing the eclipse from a high vantage point, such as on top of a hill, you will see the incoming shadow racing across the ground towards you.

Totality is seconds away. Until this point, the partial eclipse (over 99 per cent) has not been safe to look at directly – you're still using your eclipse glasses (or similar). In a few seconds that will change, but don't be too hasty. As the edge of the Moon finally blocks the edge of the Sun, there are still beads of light visible along the Moon's edge. These are called Baily's beads, named after English astronomer Francis Baily (1774–1844) who first reported these beads of sunlight during the 1836 total solar eclipse. But what are they? The Moon's surface is littered with millions of craters. A fraction of them can be seen as blemishes along the circular edge of the Moon in the sky. In the seconds after the Moon's edge 'touches' the Sun's edge, sunlight continues to shine through them, giving a bead-like appearance along the otherwise fully eclipsed Sun. Although we primarily see the same side of the Moon from Earth, it does 'wobble' throughout its orbit – an effect called libration – and as a result, what we see at the edge of the Moon can change by 7 degrees of longitude across its surface. This apparent wobble of the Moon, combined with

Baily's beads, seconds before totality.

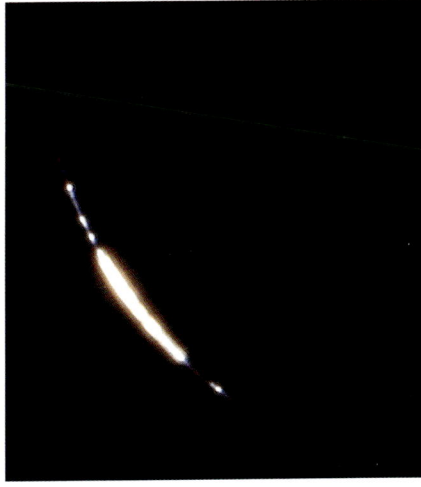

The diamond ring
provides the last
glimpse of sunlight
before totality.

the changing relative angles between the Sun and Moon from eclipse to eclipse, means that Baily's beads are close to unique for every total solar eclipse.

As the sunlight quickly disappears from each Baily's bead (in a matter of seconds), there comes a point where the last bead of sunlight shines through the final visible crater on the Moon's edge. This final stage is referred to as the diamond ring (after its appearance in photographs). It only lasts for a fraction of a second before vanishing. When it does, it is time to enjoy the totality phase of the total solar eclipse without your eclipse glasses.

TOTALITY

The main event has begun – totality is here! The Sun is now fully eclipsed by the Moon, and the faint corona is visible to the naked eye. The moment totality begins is called 'second contact', as the edge of the Moon touches the edge of the Sun for the second time.

Beyond astronomy, beyond science, a total solar eclipse is one of the most beautiful things we can witness. If anybody is ever 'unsure', or 'can't remember' if they've ever seen a total solar eclipse – then they haven't seen one. The totality phase is truly unforgettable. There is a common misconception that a 95 per cent partial eclipse is 95 per cent the experience of a total solar eclipse, but this is false. When it comes to totality, you either see it, or you don't. The difference between a total solar eclipse and partial solar eclipse is the difference between night and day. Partial eclipses are certainly exciting, but if the thrill of a partial eclipse were akin to taking a ride in a plane, then that of totality is jumping out of the plane.

You are now in the shadow of the Moon. Just as, earlier, the incoming shadow approached from the west, facing east now provides a view of the blue sky rushing away from you towards the horizon – this is called the racing shadow.

Let's return our attention to the corona. The corona is the hot atmosphere of the Sun, about 1 million degrees C (1.9 million degrees F) in temperature. Like the rest of the Sun, it is comprised of plasma made up of primarily hydrogen, helium and free electrons. At a glance, the corona looks like a luminous white halo engulfing the jet-black Moon. But if you take a moment, you can make out its finer structure. The brightest part is most intense closest to the Sun, where the density of electrons is highest. It is fainter at larger distances, but on a clear day you can see detail in the corona at least two to three times further from the Sun than the size of the Sun itself. Many interesting structures may be visible in the corona, but they vary from eclipse to eclipse. They can all be seen by the human eye, but may be better captured with photography. Physically, these structures are dictated by the magnetic configuration of the corona. The image on page 97 labels some of the features you may spot within the corona. They are listed on the next page.

' At a glance, the corona looks like a luminous white halo engulfing the jet-black Moon. '

CORONAL STRUCTURES

1. **Coronal loops:** Loops in the lower and middle corona, arching out and back into the Sun. These are magnetic loops filled with denser plasma, usually found above groups of sunspots.

2. **Helmet streamers/coronal streamers:** Streams of plasma emanating away from the Sun. Their structure resembles a candle flame at the base, they then point outwards with a long, narrow tip. If there are several of them the corona may look similar to a five-pointed star.

3. **Polar plumes:** Faint streaks at the Sun's north and south pole, directed outwards from the Sun. These are formed by plasma following magnetic fields that connect with one another at the edge of the Solar System.

4. **Coronal mass ejections (CMEs):** Eruptions of plasma from the solar corona, travelling at speeds ranging from 400 km per second (or 900,000 miles per hour) to 3,000 km per second (or 6.7 million miles per hour). Although this seems fast, it still takes them several hours to travel away from the Sun. If you're lucky, you may see these in the solar corona during totality. Although they are moving quickly, you won't notice them change position during the few minutes of totality.

In addition to the faint white structures in the corona, you may spot brighter red features at very low heights above the Moon's edge. This is the chromosphere, a thin layer of the solar atmosphere, above the surface (the photosphere), but beneath the expansive corona. It is cooler than the corona, and has a thickness of only 1 per cent of the Sun's radius. The chromosphere has a strong emission in a specific wavelength

Polar plume

Helmet streamer

Active region loops

Helmet streamer

Polar plume

Coronal features
visible during a total
solar eclipse.

of light called H-alpha (see page 84), which is a red wavelength
of light produced by hydrogen. Because of this emission, the
chromosphere appears red to the human eye. During totality,
you may see the following chromospheric features, which are
also outlined in the image above.

1. **Chromospheric beads:** Like Baily's beads seen just before totality (see page 93), these are beams of photospheric light passing through craters along the Moon's edge to create a bead-like appearance that can be seen during totality. The chromosphere is only around 1,000 km (600 miles) wide, but the red emission can be seen through craters on the Moon's edge at the start of totality on one side of the Moon, and on the other side towards the end of totality.

2. **Chromosphere layer:** In addition to chromospheric beads, totality also grants us a brief look at the chromosphere itself. It is seen as a thin band of red around the edge of the Moon.

3. **Prominences:** These magnetic arches of chromosphere material are suspended thousands to tens of thousands of kilometres above the chromosphere. (Prominences are also called 'filaments' when viewed against the Sun, but they are the same thing.) Although different and spatially distinct from the chromosphere, they are made of the same plasma, so appear red to the human eye. These loops are smaller than coronal loops (see page 96). Their size varies significantly, ranging from barely noticeable bumps to almost-unbelievable structures around the dark Moon. Like CMEs (see page 96), filaments can also erupt, so if you're lucky, you may see them mid-eruption during totality, seemingly suspended above the Sun and Moon. For the few minutes of

Chromospheric features visible during a total solar eclipse.

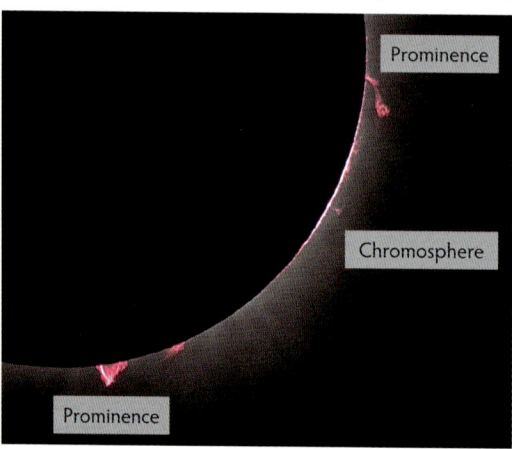

totality, prominence eruptions will appear stationary to our eyes.

4. **Solar flares:** The explosive conversion of magnetic energy in the Sun's atmosphere into the heating of plasma, acceleration of particles, and light. They appear differently at different wavelengths, but exhibit strong H-alpha emission (see page 84) during

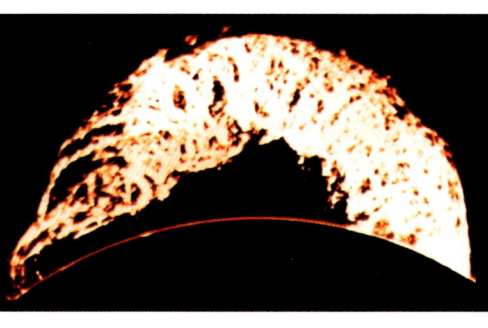

Coronagraph observations revealing a large prominence on the Sun. You are lucky if you spot one during a total solar eclipse.

the later phase of the flare. At this wavelength, solar flares appear as red loops above the edge of the Moon. They look very similar to filaments, except, unlike filaments, flare loops do evolve and grow over a few minutes. If you're lucky enough to see a solar flare during a totality of 2 minutes or more, you may notice this changing structure. Be careful though, as the Moon slowly moving across the Sun can give the appearance that a prominence on one side of the Sun/Moon is growing (and on the other side, shrinking), when in reality they remain a constant size. This can lead to the easy misidentification of a prominence as a solar flare.

The duration of the totality phase of a total solar eclipse varies from event to event. The shortest periods of totality last for a few tens of seconds (less than a minute), but the longest events can last 7 minutes or more. If you have a short period of totality, you may not have time to appreciate anything beyond the appearance of the corona (and any chromosphere features) above you. If you have some more time during totality, then there are some further phenomena you should look out for.

Take a look around and appreciate the colours above you. The Moon, still blocking the Sun, will appear to be a dark, velvety black. The corona around the Moon will have a

Strange sky colours
during a total solar
eclipse.

white-silvery appearance against a dark sky, but the whole sky
will still have a faint silver glow. You may see a colour gradient
across the sky above you in the direction of the eclipse path,
giving a strangely coloured feeling of dawn or dusk. If totality is
long at your location, the sky will become dark enough for stars
and planets to appear. Have a look over the whole sky, as you
may even see constellations not typically visible at the time of
year. The strange behaviour of animals and wildlife around you,
first noticeable during the deep partial phase, will climax too.

And finally, turn your attention to the people in your group.
How are they reacting? Screams of joy? Stunned silence?
Everyone acts differently during a total solar eclipse, especially
if it is their first. Set up a camera and point it at your group
prior to totality. Your reactions will be fun to watch back later.

The totality phase is the highlight of what all eclipses have to
offer. It ends at third contact, when all the previous steps happen
again in reverse. There's another chance to see the racing shadow,
diamond ring, Baily's beads and shadow bands, followed by the
slow reversal of the partial eclipse phases. If you're in a crowded
area, most people will rush to leave as soon as totality ends.
However, if you have the time, stay to watch the partial phases
again. The partial eclipse ends at last contact, as the final glimpse
of the New Moon vanishes from view – leaving a view of the Sun
that could easily convince you nothing happened at all.

SOLAR CYCLE VARIATIONS

The solar corona is not identical for every eclipse. Although it may not seem so on an ordinary sunny day, the Sun is incredibly dynamic – continuously evolving over timescales from hours to decades. The Sun follows an 11-year cycle of solar activity. At the peak of this cycle, called solar maximum, activity on the Sun is high, with frequent sunspots, solar flares (see page 99) and CMEs (see page 96). At the bottom of this cycle, called solar minimum, months can pass without a single sunspot, solar flare or CME. The corona looks unique during every solar eclipse, but follows the pattern of the solar cycle. During solar maximum, for example, there will be many sunspots on the Sun, forming a corona comprised mainly of multiple coronal loops and helmet streamers (see page 96), giving the look of a multi-point star. Prominences (see page 98), solar flares and CMEs are also more common during an active period of the solar cycle. During solar minimum, on the other hand, the magnetic field of the Sun is simpler. The outcome of this is a corona dominated by two polar plumes, and a single helmet streamer on either edge of the Sun. Although the corona appears more basic during solar minimum, it is less magnetically confined than during solar maximum, so will appear larger in the sky.

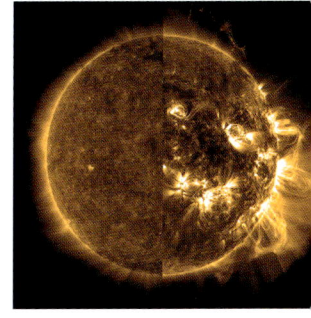

A comparison of the Sun at solar minimum (*left*) and solar maximum (*right*), observed in extreme ultraviolet light.

A comparison of a total solar eclipse between solar minimum (*left*) and solar maximum (*right*).

ANNULARITY

Annular solar eclipses are a special type of partial solar eclipse. The relative size of the Moon in the sky is not constant (see page 55) due to the slight eccentricity of the Moon's orbit around Earth, which changes the distance between the Earth and Moon from 357,000 km (222,000 miles) at its closest approach, to 407,000 km (253,000 miles) at its farthest. If a solar eclipse takes place while the Moon is close to its farthest distance from Earth, it is not big enough to block out the Sun. Instead, as the Moon passes in front of the Sun, it forms a 'ring of fire' encircling the Moon, known as an annular solar eclipse. Its peak, when a fully connected ring of sunlight surrounds the Moon, is called annularity. The experience does not compare to totality during a total solar eclipse because the corona (and other phenomena associated with totality) never becomes visible. However, an annular eclipse offers all the same points of interest visible during partial eclipse phases.

A key difference between an annular solar eclipse and strong partial phase of a total solar eclipse is the shape of the crescent Sun. Before and after totality in a total solar eclipse, the crescent shape of the Sun is short. Contrastingly, moments before annularity, the smaller Moon forms a long 'devil-horn'-shape crescent Sun, slowly encircling the entire Sun until the two ends of the crescent meet and form the full ring of fire. Annular solar eclipses can make superb partial solar eclipse shapes.

Top: A 'devil horn' partial solar eclipse, visible just before annularity.

Centre: The annularity/'ring of fire' stage of an annular solar eclipse.

Bottom: The final partial phase of a total solar eclipse. The crescent Sun is far shorter than seen during annular solar eclipses.

six

ECLIPSE PHOTOGRAPHY

HAVING READ THIS FAR, you've seen dozens of eclipse photos and perhaps wondered what it takes to take them. Depending on how much time you're willing to invest, you could produce images that match or exceed any of the examples on the previous pages. Eclipses are among the most straightforward astronomical events to capture, and ubiquitous, powerful smartphone cameras have opened up a world of astrophotography to everyone. The most extraordinary photos demand careful consideration of equipment and image processing, but it needn't be costly. This chapter covers the essentials, which will help you get started or take your next steps. Photography and image processing are always advancing, but the advice ahead is evergreen.

THE BASICS

SCALE

Solar and lunar eclipse are visually distinct, but they share similarities that are relevant to how we approach photography. The first of these is scale. The Moon is huge, and the Sun is colossal, yet both subtend just over half a degree of angular size in the sky. At their great distances, they appear small enough to cover with the tip of a little finger at arm's length. You may have taken a photo of a beautiful Full Moon with your smartphone, only to see it represented as a tiny dot in the image – not the finely resolved disk your eyes perceive. There are plenty of beautiful ways to capture an eclipse with a wide field of view, but it will always appear small without some significant magnification.

> ' There are plenty of beautiful ways to capture an eclipse with a wide field of view, but it will always appear small without some significant magnification. '

MOVEMENT

As objects in the sky, the Sun and Moon aren't stationary, even if they appear so. Sidereal motion from the rotation of the Earth, combined with the gradual change in the position of the Moon, must be considered when using very long focal lengths. Ultimately, a sturdy tripod will suffice for most images, but for total solar eclipses you may wish to consider a star tracker – a device that compensates for the rotation of the Earth. This will make high dynamic range (HDR) compositing (see pages 118–122) easier to achieve.

Either side of totality, dazzling Baily's beads wash out surrounding details. In this single shot, the camera sensor's excellent dynamic range allows faint pink prominences to be recovered even as the unfiltered light of the Sun spills through valleys on the lunar limb.

DYNAMIC RANGE

Central eclipses of both the Sun and Moon (see page 54) are dark events, occurring either during the night or creating a form of local night. Modern digital cameras are very capable in low-light conditions, but you should also consider the dynamic range – that is, the range between the deep shadows and brilliant highlights of the scene. Partial lunar eclipses are much darker in the umbral region than the penumbral region (see page 60). Total solar eclipses involve exceptionally bright Baily's beads (see page 93) alongside much fainter prominences (limb filaments) and the still more elusive corona (see page 25). The result can be improved with HDR processing, but to be sure you are maximizing dynamic range, always capture photos using your camera's raw format. Compressed formats, such as JPEG, discard crucial information and make it impossible to recover certain details.

EQUIPMENT

Virtually all modern cameras are suitable for capturing eclipses. Familiarize yourself with your camera's manual mode, as you'll need to control all aspects of the capture process, from the exposure time to the sensitivity (ISO) setting. A camera with built-in interval shooting modes and bracketed exposure functions is a bonus. Otherwise, it's worth investing in an inexpensive

Even with a 600 mm focal length, the Moon doesn't fill a 35 mm format full-frame mirrorless sensor, but lenses in this range can be used to frame eclipses close to the horizon with a foreground.

←

Using a telescope with a focal length of about 1 m (3 ft) will provide an excellent image scale for high-resolution eclipse photography.

intervalometer (or remote shutter) so you can expose photos in a sequence, without touching the camera.

Telephoto lenses increase the image scale, although the typical range may fall short of your expectations. If you have an astronomical telescope, it can serve as an excellent eclipse lens with a suitable adapter. DSLR and mirrorless cameras can be attached to most telescopes using a T-adapter and corresponding T-ring. Check that your telescope will focus using this method first; this is known as prime focus photography. If you find that the camera doesn't push in far enough to achieve focus, a Barlow lens may solve the problem. This component sits between the telescope and the camera, and amplifies the image scale (usually by a factor of two). A Barlow lens is similar to a tele-extender for a photographic lens. It's an excellent way to increase the magnifying power of a compact telescope and to assemble a portable eclipse photography kit.

Virtually all modern cameras are suitable for capturing eclipses.

Your camera's live view function will help you to focus and correctly expose your shot. Use the zoom function and focus manually on the limb or the lunar or (filtered) solar disk. If you're using a telescope, the image will be flipped in one or both axes, and you will need to rotate it yourself if you would like it to show the orientation that you see with your eyes.

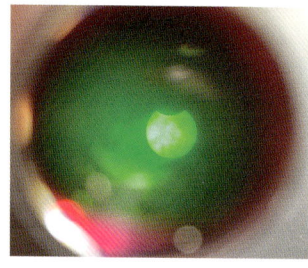

A partial solar eclipse is seen by an iPhone camera through the eyepiece of a small refracting telescope. The image is green due to the continuum filter that provides enhanced contrast in combination with a white-light filter.

Various eclipse photos captured using iPhone telephoto lenses. Details are very limited, but the essence of the eclipse is nevertheless represented. Smartphones will no doubt continue to improve, but their resolving power will always be limited by small optics.

USING SMARTPHONES

Smartphone cameras have come a long way since their introduction. Modern flagship phones provide excellent low-light performance, although in many cases the result is achieved using computational photography. In such cases, the smartphone makes decisions to try to improve the final image. This isn't ideal for all applications, and you may want to enable manual settings and try to capture images in raw formats for later processing.

Cleverly designed compact telephoto optics allow some smartphones to produce remarkably detailed images of the Moon, although the image scale and effective resolution will always be limited. Nevertheless, if a smartphone is all you have for now, it's well worth trying to take eclipse photos with it.

Your smartphone can also be paired with a telescope or even a pair of binoculars by simply holding it close to the eyepiece. This is called afocal imaging. For more stability and consistency, use a smartphone eyepiece adapter, which will hold your device in place. It takes some practice to make this technique work, but it does alleviate the need to properly pair a bulkier camera.

LUNAR ECLIPSE PHOTOGRAPHY

The Moon is generous enough to offer us an opportunity to practise capturing its full phase every 29.5 days – generally once a month. Therefore, you can practise lunar eclipse photography and determine how to approach your shots using your equipment before the next eclipse arrives. Use planetarium software such as the free programme Stellarium (stellarium.org) to determine the Moon's altitude and direction at your location, then scout for ways to make your shot more interesting.

Frame up the Moon with a unique foreground for an interesting shot. Here, a lunar eclipse appears behind the top of the Washington Monument on the National Mall, Washington, DC, USA.

Long-exposure photograph of the Milky Way.

A combined sequence of photos creatively reveals the curvature of the Earth's umbral shadow as the Moon passes through it.

Typically, photos of the Moon can be achieved in a small fraction of a second. If it enters the umbra (see page 61), it becomes significantly fainter. Depending on the speed of your optics and the ISO setting on your camera, you may need to expose for anywhere up to half a second. Ideally, exposures should be as short as reasonably possible. Scintillation (which causes stars to twinkle) will smudge the Moon's surface features during longer exposures, so try to keep the shutter speed high to preserve details. During the course of a total or fine partial lunar eclipse, the Moon will traverse a distance several times its own diameter relative to the stars. If you capture photos at regular intervals, why not merge them to reveal its progress?

The umbral shadow is sufficiently dark that long exposures can be taken, revealing surrounding stars. Very long exposures will reveal the motion of the Moon unless your camera and lens (or telescope) are compensating for the rotation of the Earth. When using a wide-angle lens, exposures of several seconds or more are possible without apparent lunar motion.

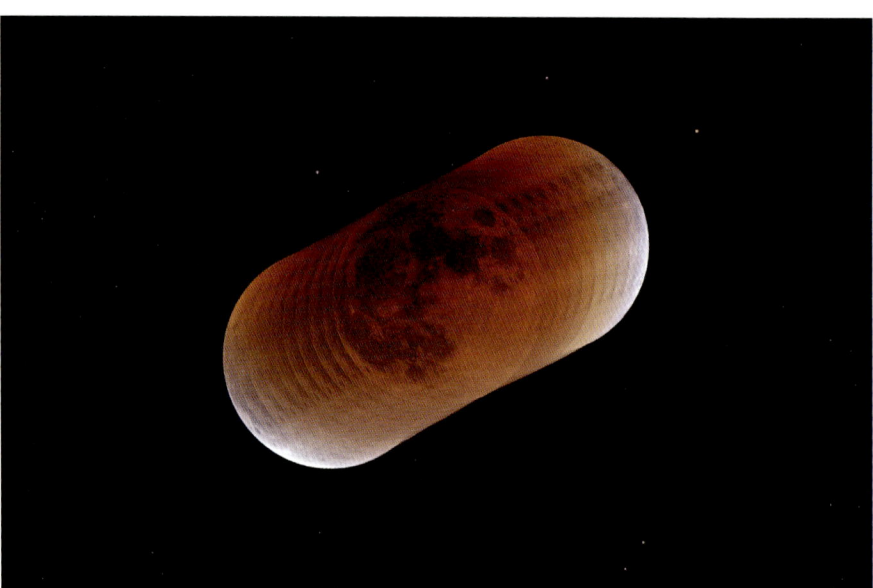

During the summer months in the northern hemisphere, the Full Moon occurs near a portion of the ecliptic that is close to the brightest region of the Milky Way. From equatorial latitudes, on a clear night it is possible to see the centre of our galaxy during a total lunar eclipse, and photographs can reveal this. You may need to slightly overexpose the eclipse in order to reveal the dust clouds that add texture to the clouds of starlight beyond, but the rich umbral colours will be preserved, resulting in an unusual photo.

Shooting in raw, you preserve shades that aren't shown on your camera's screen. Later, you can use digital processes to temper highlights and enhance shadows to create images that more closely represent what you can see with your eyes. A single photo allows you to go so far. If you want to achieve the best results, you'll need to try HDR processing. During a partial lunar eclipse, there is a very stark difference in brightness between the umbral and penumbral light on the lunar surface. If you use your phone to capture one photo of the penumbra and another of the umbra, you can combine them and enjoy the best of both worlds.

Two photos, one with a shorter exposure (*top*) and one longer (*bottom*), have been combined in Photoshop to produce an HDR composite (*centre*), which preserves the colour of the umbra and details in the penumbra. The result more accurately depicts the eclipse as seen by eye through a telescope.

You'll need to use an image processing package that supports layers and masking, such as Adobe Photoshop, or find specific HDR processing software to automate the process. In either case, the images need to be aligned. Automatic alignment can struggle, so a manual approach is recommended, but tracking the Moon using a star tracker will mitigate the need to align the images in the first place. A star tracker may offer a lunar rate, as the Moon moves eastward among the stars even as the Earth turns.

SOLAR ECLIPSE PHOTOGRAPHY

There are unique considerations when photographing a solar eclipse, just as there are when observing one. During partial eclipses, the Sun's disk is always a bright subject, unless it is obscured. When the disk is low over the horizon, while it is still not suitable for direct viewing, your unfiltered camera can be used to capture it as the atmosphere itself is the filter. Using a very fast shutter speed, consider using buildings, trees or other horizontal structures to create dramatic silhouettes.

Cloud is another natural filter that sufficiently reduces the brightness of the Sun for a camera to capture its disk. A smartphone with a telephoto lens will reveal the crescent shape of a partial eclipse and the textured scene around it. In clear weather, use a white-light filter in front of the camera lens for

 ←

The Statue of Freedom at the top of the Capitol Building in Washington, DC, USA, stands in front of this gorgeous solar crescent, accompanied by wisps of cloud.

↑

You should not look directly at a solar eclipse, even when cloud is present, but your smartphone camera can capture the moment for you.

direct photography. For a smartphone, this could be the shade on one side of your eclipse glasses.

You can use a larger piece of white-light filter material to construct a removable filter for a camera lens or telescope (for example, AstroSolar Safety Film sold by Baader Planetarium in A4 sheets). To do this, use a piece of card to make a fitted tube that slides onto the front of the optical tube and attach a piece of film inside it. Ensure there are no gaps in the film but don't pull it taut – it will work perfectly well with a few ripples in it and should not be stretched. With the filter in place, you can use your camera's live view function to focus directly on the Sun, then set up your exposure. Unless your camera's screen is very bright, you might have trouble seeing the image. You can use a piece of card as a simple shade or you can drape a towel over yourself and your camera.

White-light images will often reveal interesting regions on the Sun's fluid surface – the photosphere. Small sunspots and granulation are a sign that your lens or telescope are near perfect focus. Look carefully at the black silhouette of the Moon to see if you can make out its rugged edge.

A white-light filter will help you to stay on target during partial phases and protect your camera's sensor from the intense heat and light. Here, a star tracker, aligned to the Earth's polar axis, follows the Sun. With no visible stars, you will have to use a compass and your latitude to approximate your alignment.

The crescent Sun shows sunspots and granulation in white light. Meanwhile, the lunar limb is evidently rugged in silhouette, its mountains and valleys forming imperfections in the curvature.

It is only as the diamond ring (see page 94) fades and totality begins that it is safe for you to remove the filter, which should quickly be replaced when the diamond ring re-emerges during the egress. Otherwise, you risk damaging your camera's sensor. This can be a stressful moment, and particular care should be taken during a hybrid eclipse. Totality is short, and you will need to experiment with exposure times to capture it. We strongly advise that you preplan for the execution of the capture phase, and run a routine using an intervalometer or, better yet, an exposure bracketing tool. A sequence of five or seven photos ranging from about 1/500 seconds to 1 second (suitable for most cameras at ISO 400–800) will likely ensure you capture at least one properly exposed photo and allow you to experiment with HDR processing.

A well-balanced photograph during totality shows details around the lunar limb that have not been lost to the brightness of the corona.

Use a remote shutter or camera control software (such as Moonglow Technologies Eclipse Orchestrator) to capture images continuously while you enjoy the view. It's tempting to check the first batch, but only do so if you have time – it's not worth the anguish of missing the totality entirely if you get caught up adjusting settings. Remember, the camera screen won't provide a full impression of the detail in the photo; even the histogram has limited use. Ultimately, a classic shot of totality

reveals details in the solar corona close to the lunar limb as well as large prominences leaping out of the Sun's chromosphere. The Moon itself will be perfectly dark. In very long exposures, the Moon's near side can be seen illuminated by earthshine – sunlight reflected onto it by the Earth's atmosphere.

The dynamic range of a total solar eclipse is extreme (see page 105). Immediately before and after totality, light spilling through valleys on the Moon produces Baily's beads (see page 93), which are stunningly bright even at short exposures. They are nevertheless well worth recording, as their profile is unique to each eclipse, determined by the Moon's libration.

With adequate preparation, you can achieve a beautiful photo during totality without sacrificing your enjoyment of the eclipse, but don't be downhearted if it doesn't pan out the first time you try. A total solar eclipse is a demanding subject, and your combination of camera, lens or telescope, along with your settings, will yield different results. Whatever you end up with, keep it! You may one day return to the images with newly honed processing skills, and retrieve more than you previously thought was possible.

Above: A collage illustrating the progress to totality.

Right: If your intervalometer captures evenly spaced exposures around totality, you can assemble a collage like this to preserve the Baily's beads on either side.

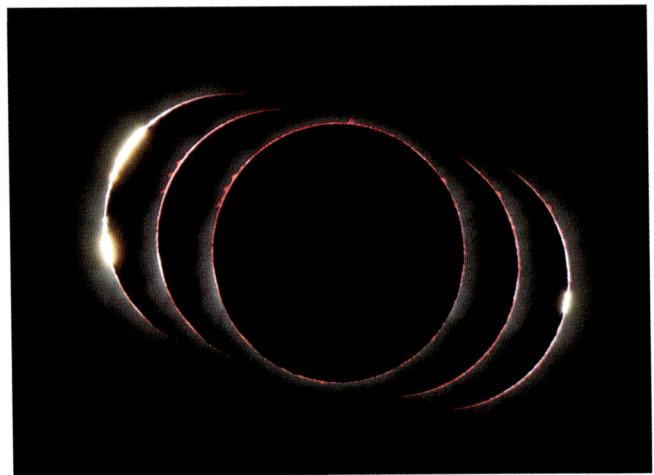

REVEALING THE CORONA

We have seen how HDR photography can enhance lunar eclipse captures. When applied to total solar eclipses, a technique using a Photoshop workflow, developed in the 1990s and first published by NASA researcher Gerald L. Pellett (1936–2021), can boost details within the corona.

Planning and shooting bracketed exposures

The foundation of HDR eclipse photography involves capturing a series of bracketed exposures that span the corona's extreme brightness range. Use autoexposure bracketing (AEB) or camera-control software to shoot between five and seven different exposures ranging from 1/500 second to 1 second. The shortest exposures capture prominences along the lunar limb, while longer exposures reveal the faint details of the outer corona. Shoot multiple bracketed sets during totality to reduce noise in the final result. Use a polar-aligned tracking mount to keep the Sun centred and simplify alignment during processing.

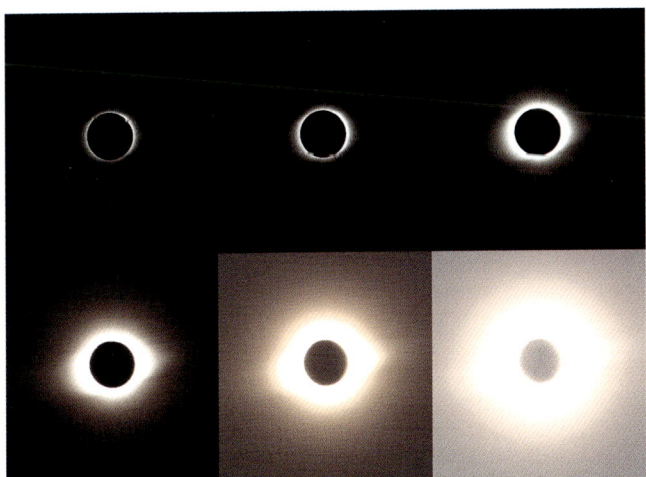

Bracketed exposures captured during the eclipse of 2 July 2019 in Chile.

PROCESSING AN HDR IMAGE

With your captures organized, you're ready to assemble them into a single HDR image to tease out coronal details.

Loading and stacking images in Photoshop

Open Adobe Photoshop and use File > Scripts > Load Files into Stack to import your bracketed exposures. The software will layer the images sequentially from shortest to longest exposure (or vice versa). If alignment is needed, select the Move tool, change the blending mode to Difference, and use arrow keys to nudge the layers into perfect alignment. The Layers window displays all your exposures stacked vertically, ready for opacity adjustments that will blend them together seamlessly.

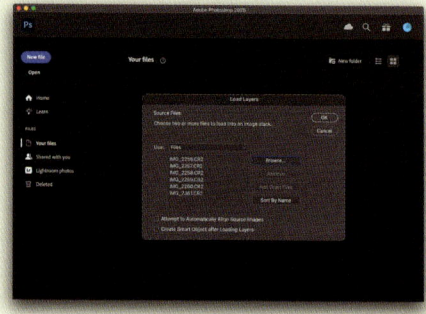

Setting layer opacity for smooth blending

Adjust each layer's opacity to create a smooth transition between exposures. For seven layers – the maximum recommended – set the opacity values as follows: bottom layer at 100 per cent, then layers 2 to 7 at 50 per cent, 33 per cent, 25 per cent, 20 per cent, 17 per cent and 14 per cent respectively. This mathematical progression ensures each layer makes the appropriate contribution to the final image without creating harsh transitions or overexposure artefacts. The result should show details from the Moon's faintly illuminated surface to prominences, coronal streamers and field stars. Save this as 'STACK.tif' after flattening the layers.

Creating a natural-looking result

For a natural eclipse appearance, apply the Camera Raw Filter (Filter > Camera Raw Filter) to your stacked image. Set both Texture and Clarity sliders to 100 and Shadows to 50 to enhance coronal features while maintaining realism. Adjust the colour balance to match your visual memory of the eclipse. This processing closely mimics what the naked eye sees during totality, revealing inner corona features and the Moon's surface details while maintaining the ethereal quality of the corona's outer regions.

Generating a blurred reference image

Open your stacked image (STACK.tif) and create a blurred version using Filter > Blur > Gaussian Blur with a radius of 3 to 4 pixels. The larger radius value increases contrast in the final result but may suppress the smallest coronal features. This blurred image serves as a reference for identifying fine details through subtraction. Save this as 'STACK_BLUR.tif' and keep it open for the next step. For optimal results, the blur radius should be adjusted based on the focal length of your telescope.

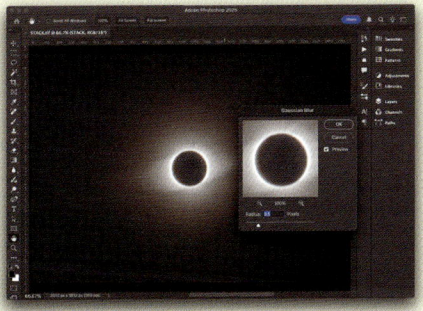

Creating a detail map through subtraction

With both the original and blurred images open, use Image > Apply Image to subtract the blurred version from the original. Set the Source to your blurred image, change Blending mode to Subtract, and increase the offset to 128. This creates a low-contrast map showing small-scale

differences between the images. The result appears grey with a dark ring where the Moon's limb is located but contains crucial information about coronal loops and streamers that will be enhanced in subsequent steps.

Compressing dynamic range with levels

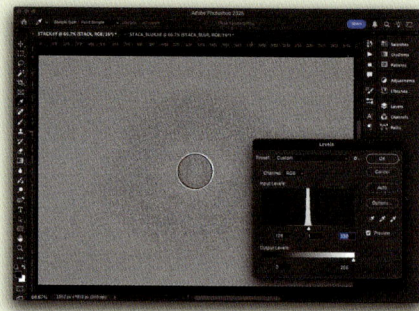

Open the Levels tool (Image > Adjustments > Levels) to compress the dynamic range of your subtracted image. Move the low value slider to 126 and the high value slider to 130, then click OK. This dramatic compression transforms the grey detail map into a high-contrast image showing coronal loop edges, streamers and background stars, though with considerable noise. Save this processed detail map as 'STACK_DETAIL.tif'. This file contains the enhanced detail information that will be applied to your original stack.

Applying the detail map to enhance corona features

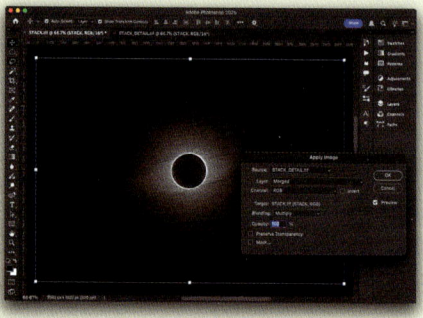

Open your original stacked image (STACK.tif) and use Image > Apply Image to blend the detail map. Set the Source to your subtracted image (STACK_DETAIL.tif) and change the blending mode to Multiply. This immediately reveals high-contrast coronal loops and fine structures, though the result will be very noisy and potentially a little 'overcooked'. Reduce the Opacity to about 75 per cent before applying the detail map to maintain a more natural appearance if this is the case. The Multiply blending mode enhances the contrast of coronal features while preserving the overall brightness structure of the corona.

Noise reduction and optional iterative stacking

Apply noise reduction using the Camera Raw Filter; if necessary, drop the Texture slider to −50 to reduce the harsh noise introduced by the aggressive sharpening process. For superior results, repeat this entire process with multiple bracketed sets shot during totality, then stack the processed results using the same opacity method. Multiple sets provide cleaner

details without requiring aggressive noise reduction, though the Moon may appear slightly smeared due to its movement across the solar disk during the eclipse sequence.

Optional: add the earthshine on the Moon

Using one of your bright photos of the eclipse, mask out the Moon's nightside with the circular selection tool. Make sure your selection is larger than the Moon, at least 15 pixels from the edges, and set the feather to 10 pixels. Copy and paste this selection onto your working image (STACK.tif) and align it. Then adjust its

exposure and levels to approximately match the darker regions of the blue sky around the eclipse. Major features on the Moon should be visible as they're illuminated by earthshine.

Final adjustments

Open the Camera Raw Filter and make final adjustments to your image. Here you can increase contrast and texture to enhance the visibility of the details in the corona. You may wish to utilize the built-in noise reduction slider or third-party noise reduction plug-ins at this stage to produce a smoother result. Your final adjustments are up to you. However, as a general rule, it's wise to avoid overcooking the image and keep it somewhat natural.

Your finished image will provide an ultimate record of the event, from the Moon's orientation to the polar streamers and equatorial wings of the solar corona. Any bright stars should also be represented.

There is no doubt that creative new ways to capture and process eclipse images have yet to be discovered, but starting with the advice in this chapter, you'll be on your way to replicating the photographs taken by the authors for this book.

The final, adjusted image reveals both details in the corona and earthshine on the Moon's nightside.

seven

UPCOMING ECLIPSES

THIS CHAPTER COLLATES the 48 lunar and solar eclipses that will occur between the first publication of this book and the end of 2036. There are maps for the 24 that qualify as central eclipses – either total lunar eclipses, total solar eclipses or annular solar eclipses. The maps show the locations from where they will be visible and their predicted timings.

DATES AND TIMES

The maps opposite illustrate the umbral (U) and penumbral (P) limits of central eclipses. Lunar eclipse maps are equirectangular and centred on the prime meridian (zero longitude). The darkest region lies between the intersection of the P1 and P4 lines, where both the beginning and end of the penumbral phase (and thus the entire eclipse) are visible. Solar eclipse limits are shown on orthographic globes. The dark line shows the path of totality or annularity across the Earth's surface.

The dates provided with each map correspond to the Universal Time (UT) date at which greatest eclipse occurs. Some inevitably begin on the previous day or end on the next one, so take care when planning to view an eclipse anywhere near the International Date Line – you don't want to be a day late!

Timing also requires special attention. Principal UT timings (penumbral and umbral contacts and greatest eclipse) are provided for each central eclipse. However, eclipse calculations are made using a different time standard called Terrestrial Dynamic Time (TT). UT differs from TT by a small factor (ΔT or Delta-T), which must be included.

Timekeeping is not as simple as we would like it to be. UT is based on the Earth's rotation and therefore is crucial for ground-based observations, but the rotation rate is changing, causing UT and TT to diverge. TT is based on atomic time, which sets the length of day to 86,400 seconds. Over the century between 1902 and 2002, the ΔT between TT and UT was about 64 seconds. For a given year, it can only be estimated based on the trend until it has been measured. As such, there may be discrepancies between the calculated UT eclipse timings and those you measure. As a general rule, always observe early rather than waiting for an accurate clock to tell you when to look. Naturally, you should check local eclipse times close to the event for your chosen location and time zone.

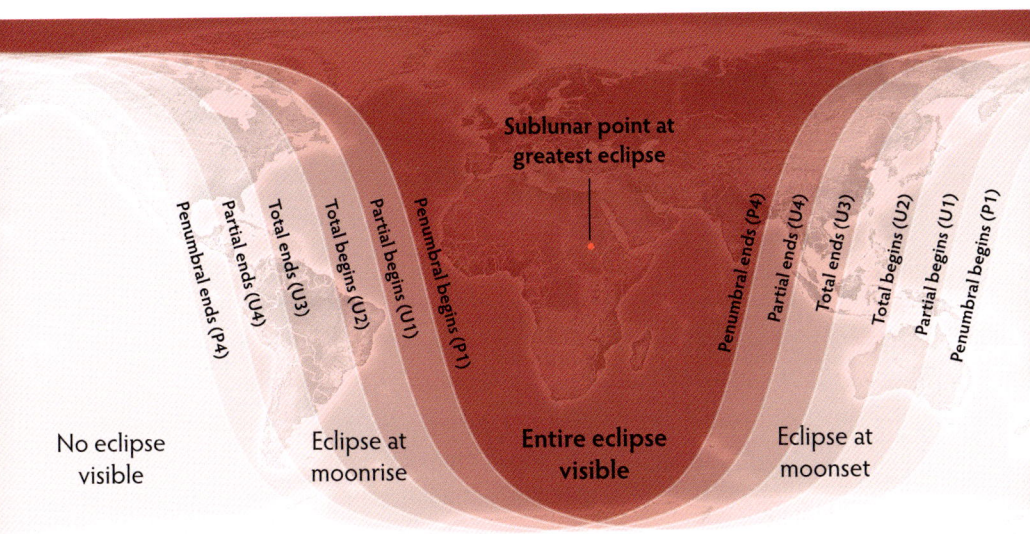

Sublunar point at greatest eclipse

Penumbral ends (P4)
Partial ends (U4)
Total ends (U3)
Total begins (U2)
Partial begins (U1)
Penumbral begins (P1)

Partial ends (U4)
Total ends (U3)
Total begins (U2)
Partial begins (U1)
Penumbral begins (P1)

No eclipse visible

Eclipse at moonrise

Entire eclipse visible

Eclipse at moonset

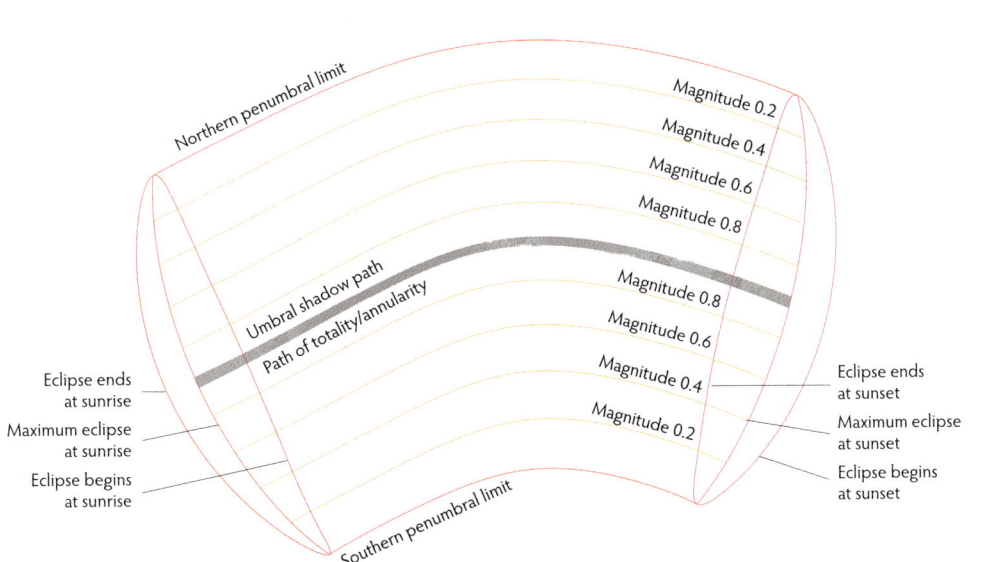

Northern penumbral limit

Magnitude 0.2
Magnitude 0.4
Magnitude 0.6
Magnitude 0.8

Umbral shadow path
Path of totality/annularity

Magnitude 0.8
Magnitude 0.6
Magnitude 0.4
Magnitude 0.2

Eclipse ends at sunrise
Maximum eclipse at sunrise
Eclipse begins at sunrise

Eclipse ends at sunset
Maximum eclipse at sunset
Eclipse begins at sunset

Southern penumbral limit

Total Lunar Eclipses
31 December 2028
26 June 2029
20 December 2029
25 April 2032
18 October 2032
14 April 2033
8 October 2033
11 February 2036
6 August 2036

Annular Solar Eclipses
6 February 2027
26 January 2028
1 June 2030
21 May 2031
9 May 2032
12 September 2034
9 March 2035

Total Solar Eclipses
12 August 2026
2 August 2027
22 July 2028
25 November 2030
14 November 2031*
30 March 2033
20 March 2034
2 September 2035

*Hybrid eclipse

The predicted timings are included for completeness, and the authors would like to acknowledge the diligent American astrophysicist Fred Espenak (1953–2025), whose work with NASA's Goddard Space Flight Center made it possible to provide this data and to generate the eclipse paths shown on the following maps.

LOCATIONS

For central solar eclipses, we've included a list of countries or territories where totality or annularity can be observed. In such cases, the central track falls somewhere within the country's borders, meaning that if you live there you can travel to see it without making an international journey. However, this does not necessarily mean the central eclipse is widely visible in that country.

CENTRAL ECLIPSES

The following pages detail 24 central eclipses that will take place before the end of 2036. There are nine total lunar eclipses, seven annular solar eclipses and eight total solar eclipses. On 14 November 2031, a hybrid solar eclipse will occur almost entirely over the Pacific Ocean. The end of its central track brings an annular solar eclipse to Panama but will briefly become a total solar eclipse over the ocean and is therefore classified as one here.

Use these maps as a guide to seek further information, particularly about safe and weather-favourable locations, as well as precise timings within your time zone as calculated close to the event.

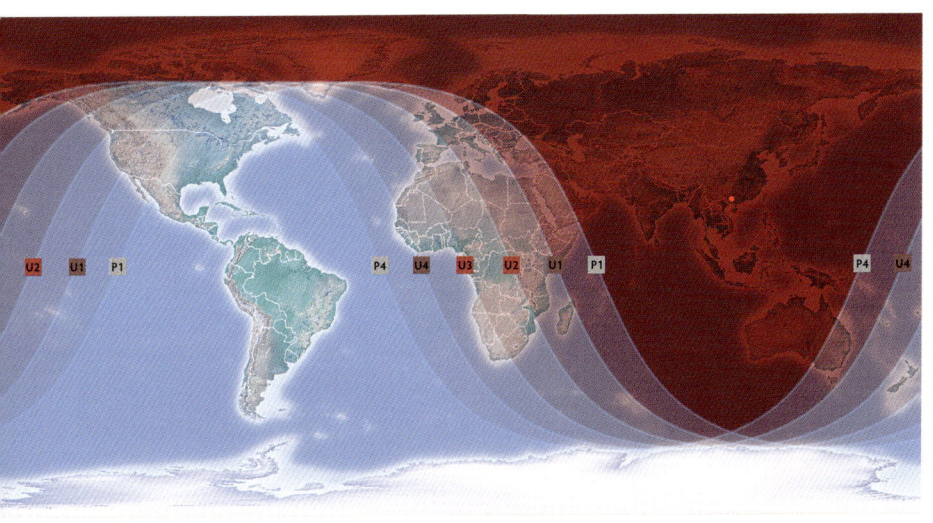

Saros series (number): 125 (49 of 72)			Total duration: 1h 11m 20s		Umbral magnitude: 1.25	
Penumbral begins (P1)	Partial begins (U1)	Total begins (U2)	Greatest eclipse	Total ends (U3)	Partial ends (U4)	Penumbral ends (P4)
12:03:49 UT	15:07:35 UT	16:16:19 UT	16:51:58 UT	17:27:40 UT	18:36:24 UT	19:40:02 UT

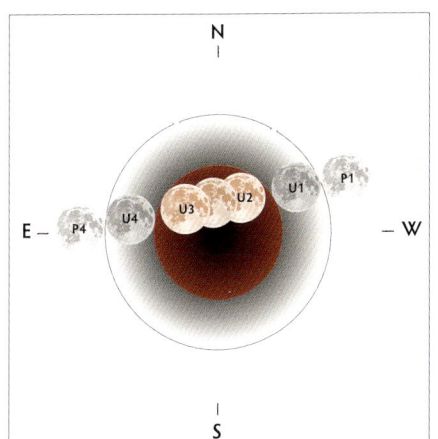

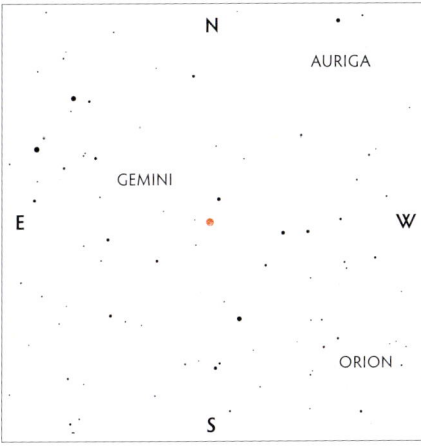

TOTAL LUNAR ECLIPSE OF 26 JUNE 2029

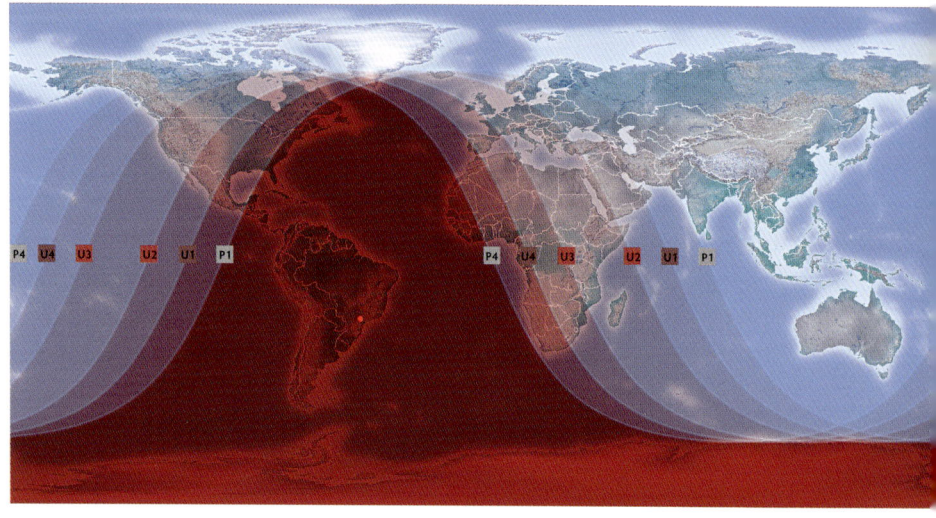

Saros series (number): 130 (35 of 72)			Total duration: 1h 41m 53s		Umbral magnitude: 1.84	
Penumbral begins (P1)	Partial begins (U1)	Total begins (U2)	Greatest eclipse	Total ends (U3)	Partial ends (U4)	Penumbral ends (P4)
00:34:34 UT	01:32:18 UT	02:31:08 UT	03:22:05 UT	04:13:01 UT	05:11:50 UT	06:09:42 UT

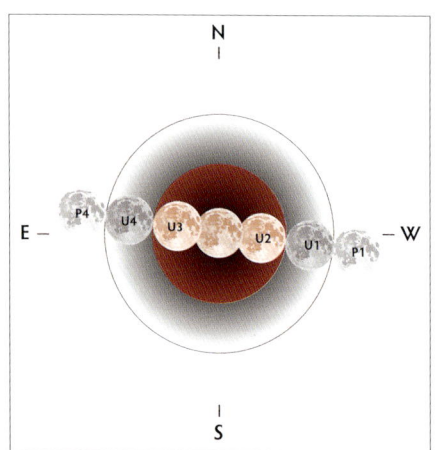

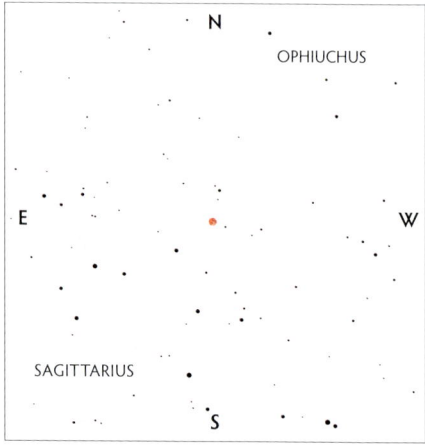

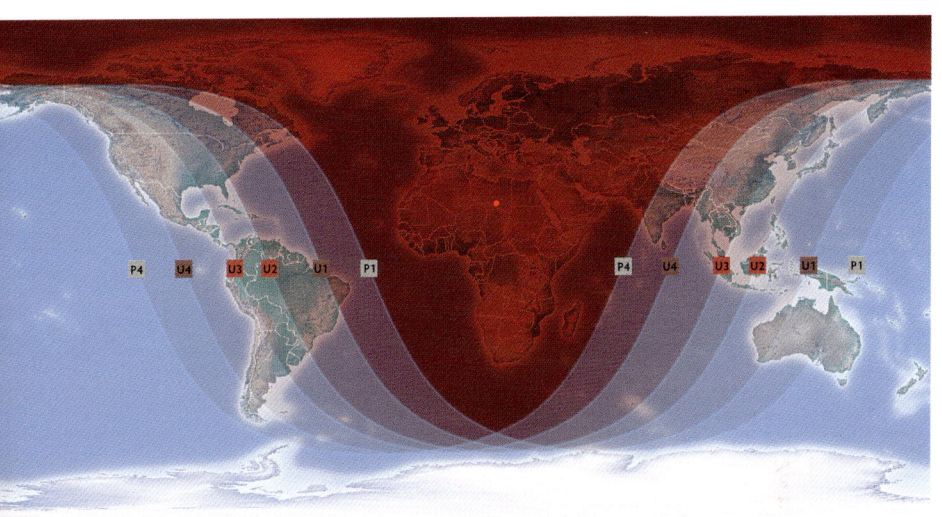

Saros series (number): 135 (24 of 71)			Total duration: 53m 40s		Umbral magnitude: 1.12	
Penumbral begins (P1)	Partial begins (U1)	Total begins (U2)	Greatest eclipse	Total ends (U3)	Partial ends (U4)	Penumbral ends (P4)
19:42:53 UT	20:55:17 UT	22:15:05 UT	22:51:53 UT	23:08:45 UT	00:28:34 UT	01:40:51 UT

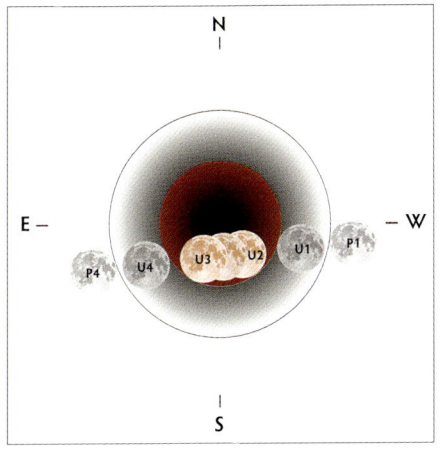

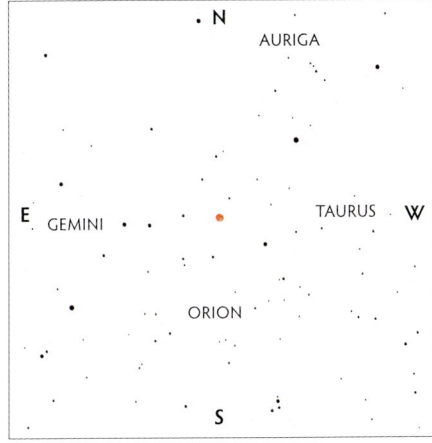

TOTAL LUNAR ECLIPSE OF 25 APRIL 2032

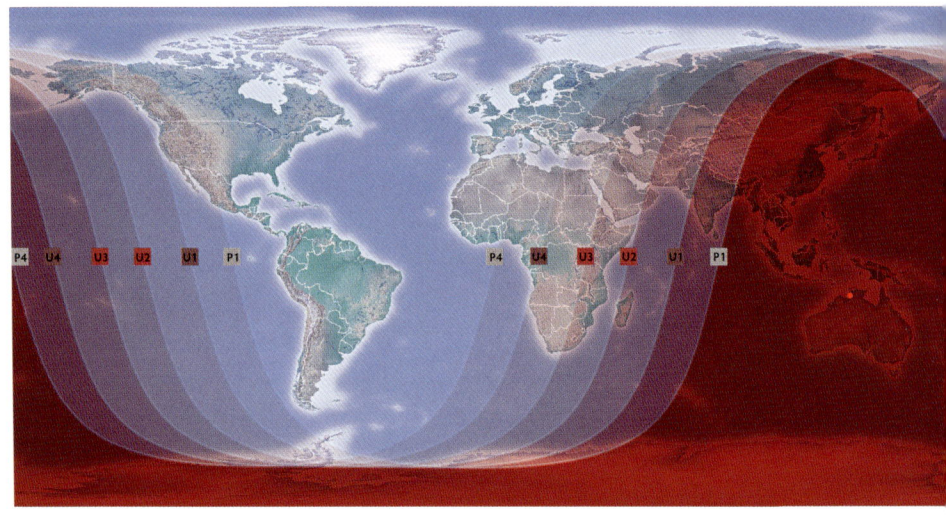

Saros series (number): 122 (57 of 75)			Total duration: 1h 5m 32s		Umbral magnitude: 1.91	
Penumbral begins (P1)	Partial begins (U1)	Total begins (U2)	Greatest eclipse	Total ends (U3)	Partial ends (U4)	Penumbral ends (P4)
12:22:16 UT	13:27:58 UT	14:40:47 UT	15:13:31 UT	15:46:19 UT	16:59:09 UT	18:04:42 UT

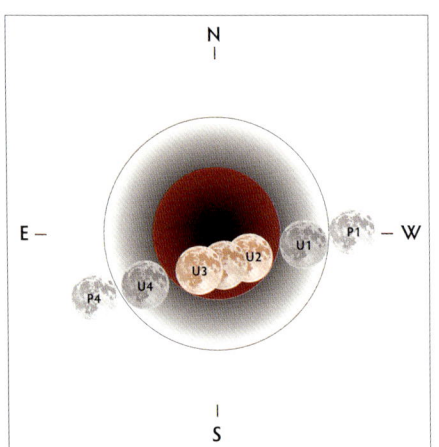

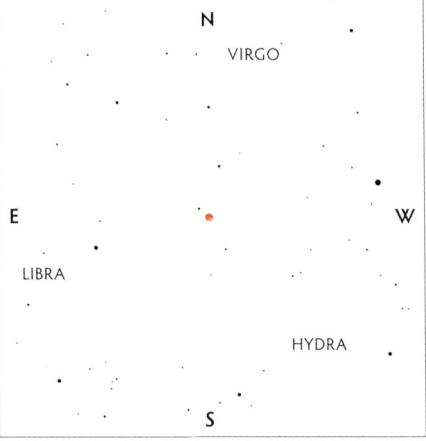

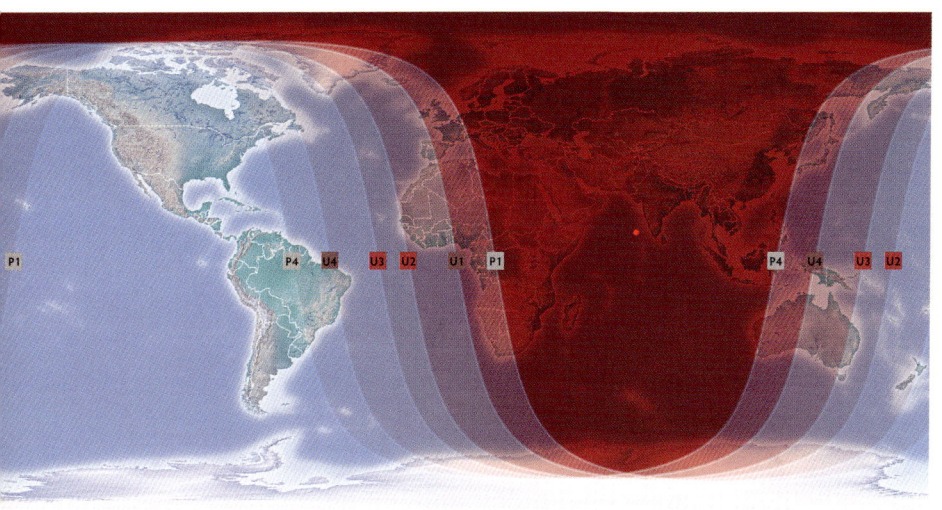

Saros series (number): 127 (43 of 72)			Total duration: 47m 7s		Umbral magnitude: 1.10	
Penumbral begins (P1)	Partial begins (U1)	Total begins (U2)	Greatest eclipse	Total ends (U3)	Partial ends (U4)	Penumbral ends (P4)
16:24:41 UT	17:24:22 UT	18:38:46 UT	19:02:20 UT	19:25:53 UT	20:40:17 UT	21:40:05 UT

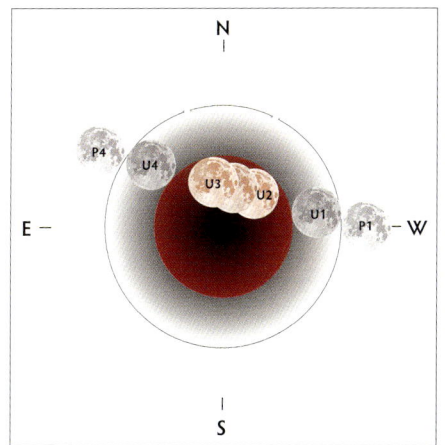

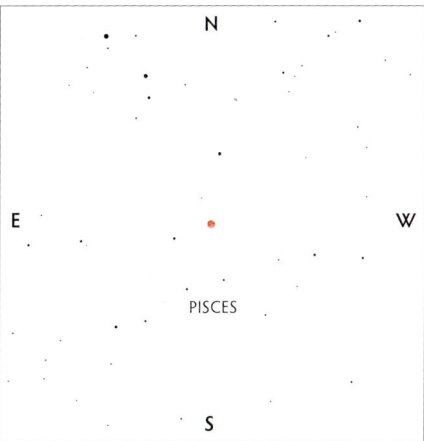

TOTAL LUNAR ECLIPSE OF 14 APRIL 2033

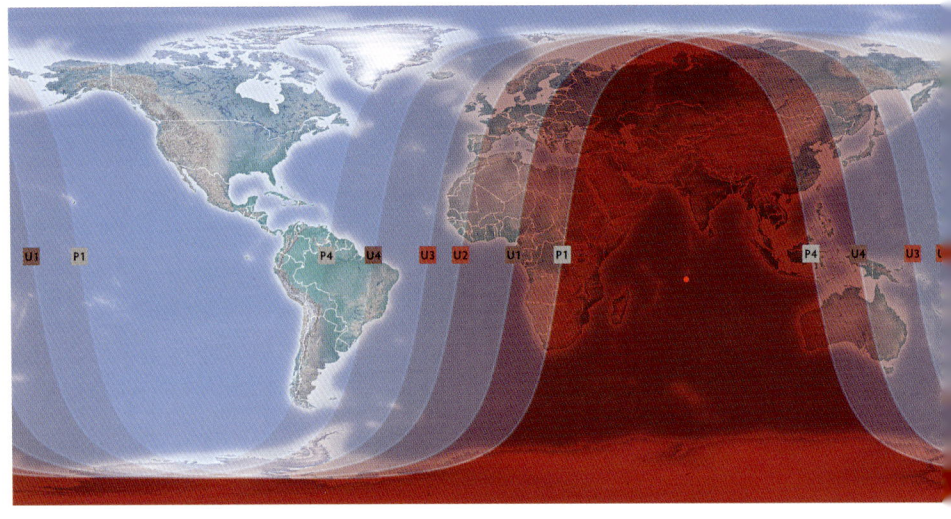

Saros series (number): 132 (31 of 71)			Total duration: 49m 12s		Umbral magnitude: 1.09	
Penumbral begins (P1)	Partial begins (U1)	Total begins (U2)	Greatest eclipse	Total ends (U3)	Partial ends (U4)	Penumbral ends (P4)
16:11:54 UT	17:25:03 UT	18:47:56 UT	19:12:31 UT	19:37:09 UT	21:00:02 UT	22:13:05 UT

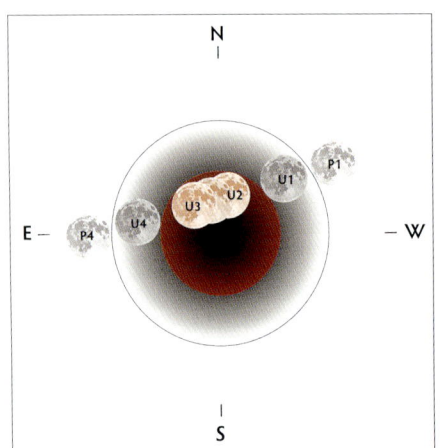

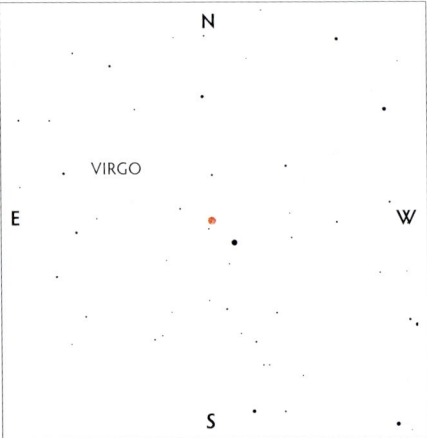

TOTAL LUNAR ECLIPSE OF 8 OCTOBER 2033

Saros series (number): 137 (29 of 81)			Total duration: 1h 18m 49s		Umbral magnitude: 1.35	
Penumbral begins (P1)	Partial begins (U1)	Total begins (U2)	Greatest eclipse	Total ends (U3)	Partial ends (U4)	Penumbral ends (P4)
08:18:44 UT	09:13:50 UT	10:15:38 UT	10:55:02 UT	11:34:27 UT	12:36:15 UT	13:31:21 UT

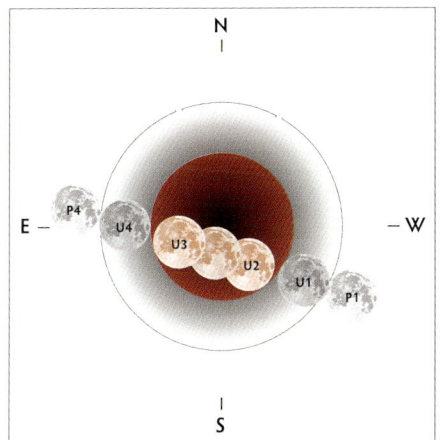

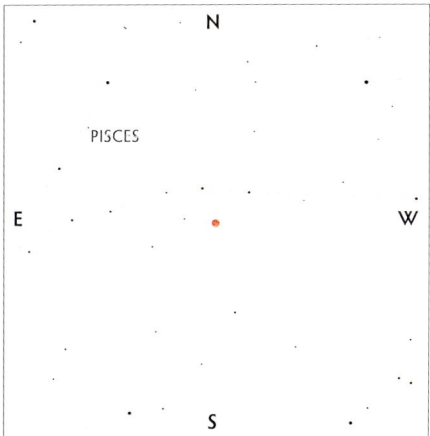

TOTAL LUNAR ECLIPSE OF 11 FEBRUARY 2036

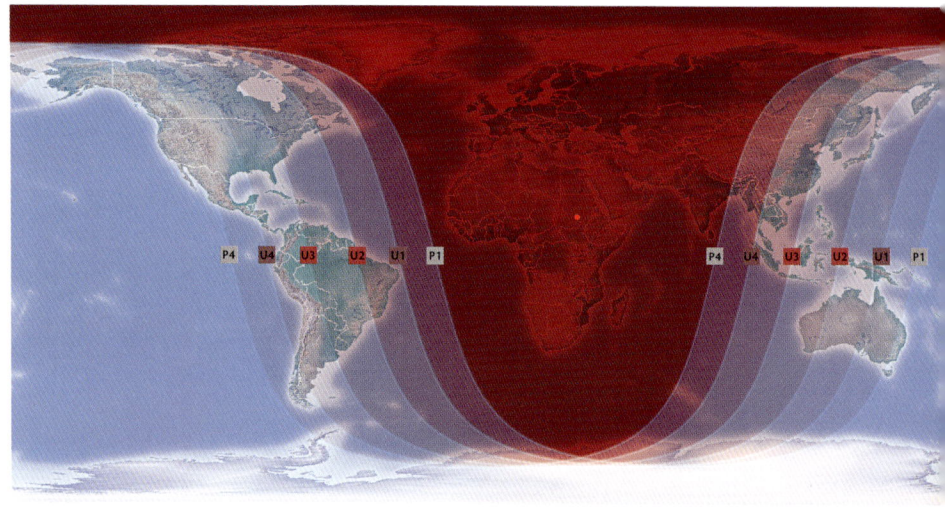

Saros series (number): 124 (50 of 74)			Total duration: 1h 14m 29s		Umbral magnitude: 1.30	
Penumbral begins (P1)	Partial begins (U1)	Total begins (U2)	Greatest eclipse	Total ends (U3)	Partial ends (U4)	Penumbral ends (P4)
19:33:44 UT	20:30:46 UT	21:34:30 UT	22:11:44 UT	22:48:58 UT	23:52:42 UT	00:49:48 UT

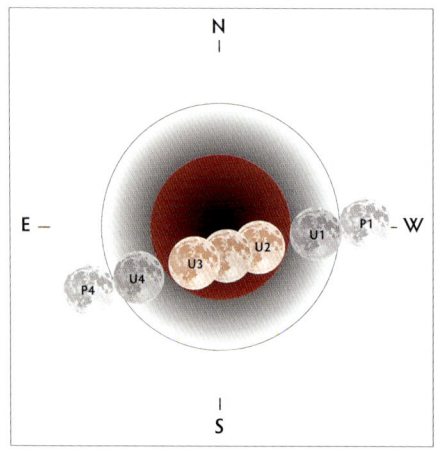

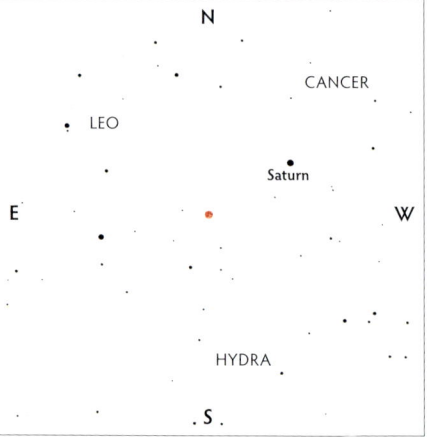

TOTAL LUNAR ECLIPSE OF 6 AUGUST 2036

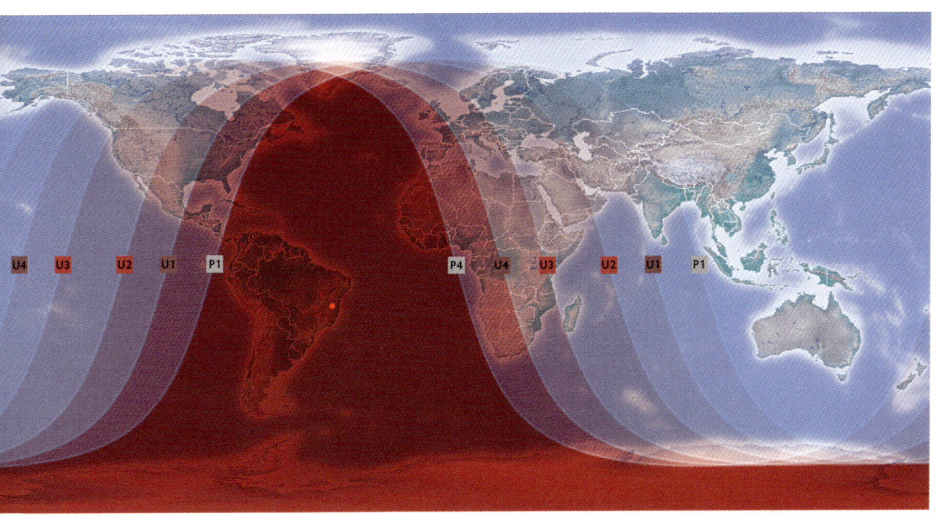

Saros series (number): 129 (39 of 71)			Total duration: 1h 35m 19s		Umbral magnitude: 1.45	
Penumbral begins (P1)	Partial begins (U1)	Total begins (U2)	Greatest eclipse	Total ends (U3)	Partial ends (U4)	Penumbral ends (P4)
23:45:07 UT	00:55:30 UT	02:03:31 UT	02:51:10 UT	03:38:50 UT	04:46:50 UT	05:57:14 UT

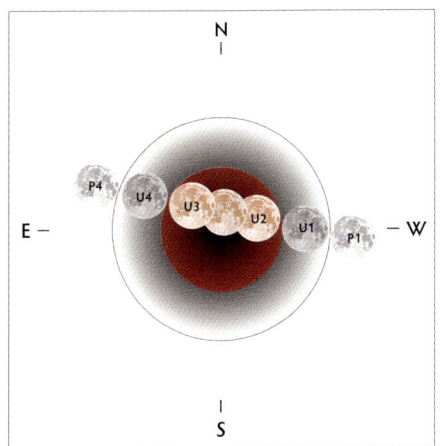

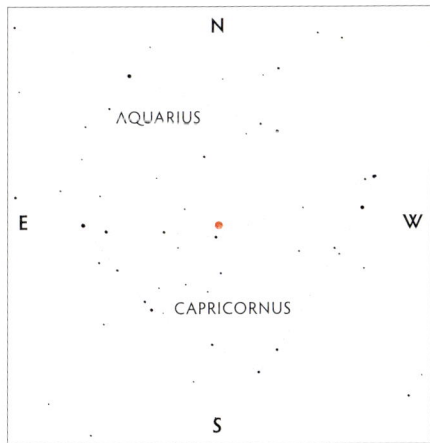

TOTAL SOLAR ECLIPSE OF 12 AUGUST 2026

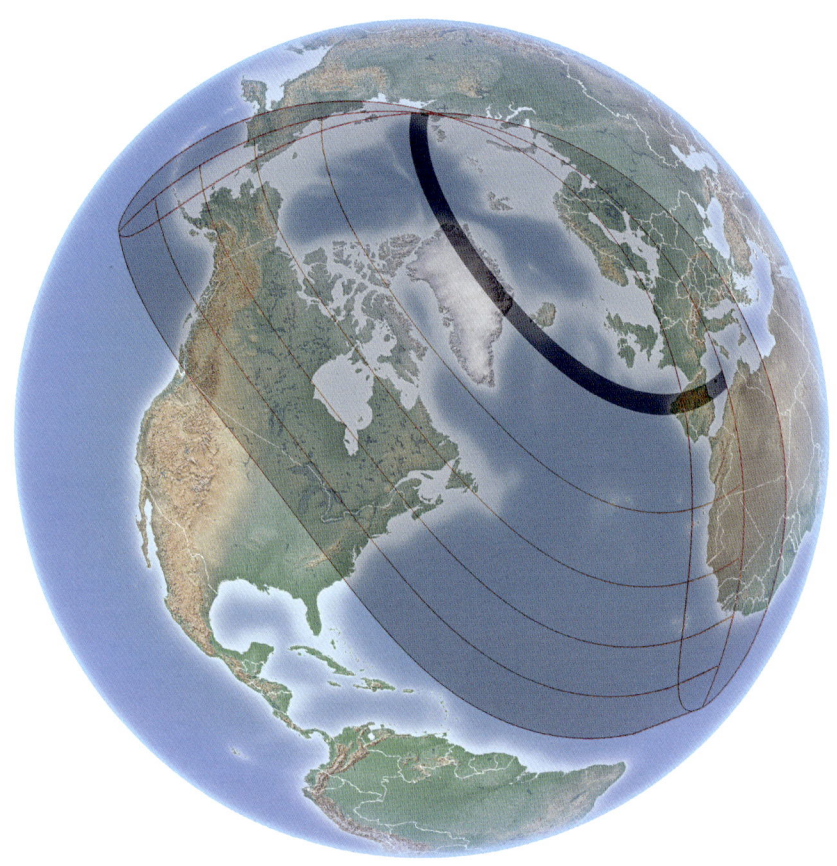

Saros series (number): 126 (48 of 72)			Magnitude: 1.04	
Greatest eclipse location: 65° 13.0' N, 025° 13.6' W			Duration: 2m 18s	
Partial begins (P1)	First totality begins (U1)	Greatest eclipse	Last totality ends (U4)	Partial ends (P4)
15:34:01 UT	16:57:54 UT	17:45:43 UT	18:33:57 UT	19:57:47 UT

Path of totality touches Greenland, Iceland, Portugal and Spain.

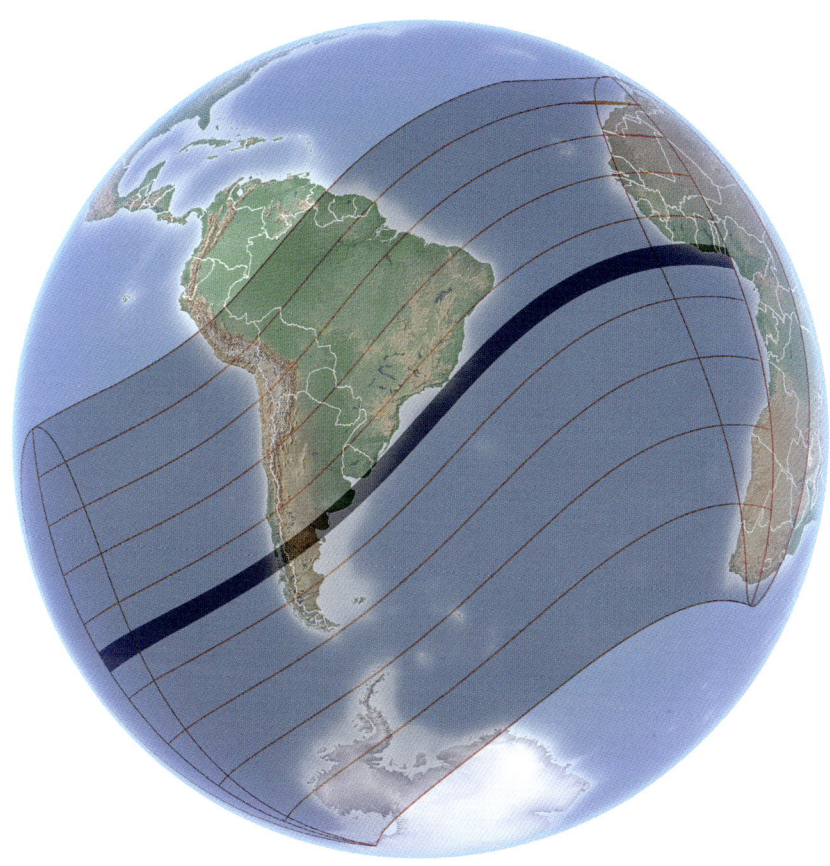

Saros series (number): 131 (52 of 71)			Magnitude: 0.93	
Greatest eclipse location: 31° 17.8' S, 048° 25.3' W			Duration: 7m 51s	
Partial begins (P1)	First annularity begins (U1)	Greatest eclipse	Last annularity ends (U4)	Partial ends (P4)
12:57:23 UT	14:03:41 UT	15:59:24 UT	17:55:13 UT	19:01:26 UT

Path of annularity touches Chile, Argentina, Uruguay, Brazil, Côte d'Ivoire, Ghana, Benin and Nigeria.

Saros series (number): 136 (38 of 71)			Magnitude: 1.08	
Greatest eclipse location: 25° 29.6' N, 033° 13.2' E			Duration: 6m 22s	
Partial begins (P1)	First totality begins (U1)	Greatest eclipse	Last totality ends (U4)	Partial ends (P4)
07:30:00 UT	08:23:16 UT	10:06:28 UT	11:49:44 UT	12:42:59 UT

Path of totality touches Spain, Gibraltar, Morocco, Algeria, Tunisia, Libya, Egypt, Sudan, Saudi Arabia, Yemen, Somalia and British Indian Ocean Territory.

Saros series (number): 141 (24 of 70)			Magnitude: 0.92	
Greatest eclipse location: 02° 58.3' S, 051° 30.5' W			Duration: 10m 27s	
Partial begins (P1)	First annularity begins (U1)	Greatest eclipse	Last annularity ends (U4)	Partial ends (P4)
12:06:28 UT	13:14:38 UT	15:07:33 UT	17:00:19 UT	18:08:34 UT

Path of annularity touches Ecuador, Peru, Colombia, Brazil, French Guiana, Portugal, Spain and Morocco.

TOTAL SOLAR ECLIPSE OF 22 JULY 2028

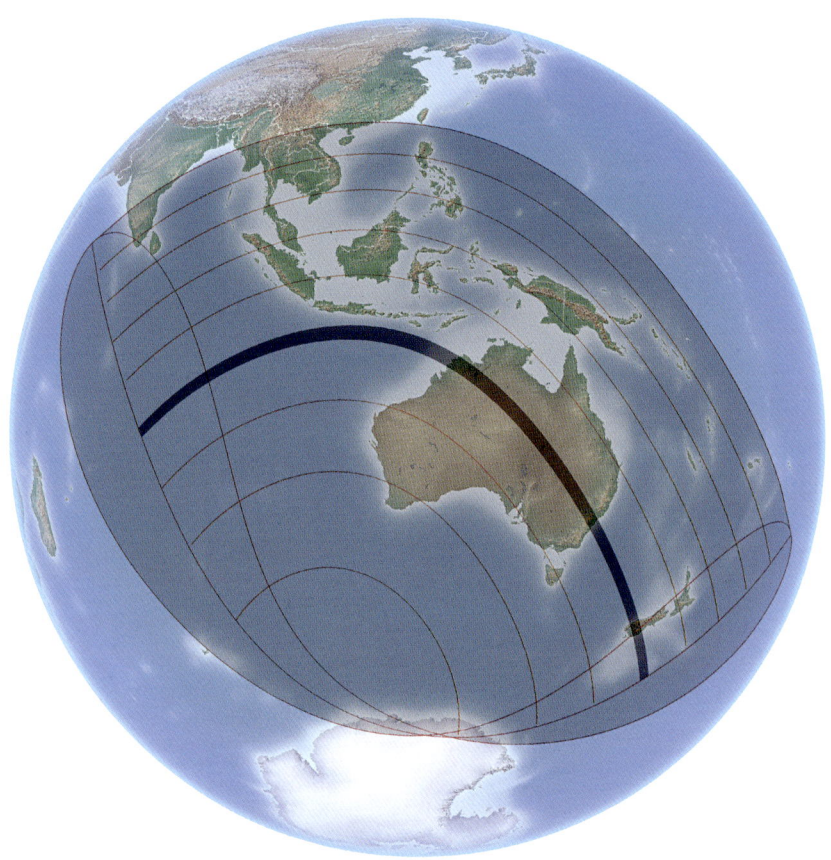

Saros series (number): 146 (28 of 76)			Magnitude: 1.06	
Greatest eclipse location: 15° 35.3' S, 126° 44.7' E			Duration: 5m 9s	
Partial begins (P1)	First totality begins (U1)	Greatest eclipse	Last totality ends (U4)	Partial ends (P4)
00:27:23 UT	01:30:30 UT	02.55:1/ UT	04:19:53 UT	05:23:00 UT

Path of totality touches Australia and New Zealand.

Saros series (number): 128 (59 of 73)			Magnitude: 0.94	
Greatest eclipse location: 56° 30.9' N, 080° 60.7' E			Duration: 5m 20s	
Partial begins (P1)	First annularity begins (U1)	Greatest eclipse	Last annularity ends (U4)	Partial ends (P4)
03:34:28 UT	04:47:00 UT	06:27:48 UT	08:08:35 UT	09:21:05 UT

Path of annularity touches Algeria, Tunisia, Libya, Malta, Greece, Turkey, Bulgaria, Ukraine, Russia, Kazakhstan, China and Japan.

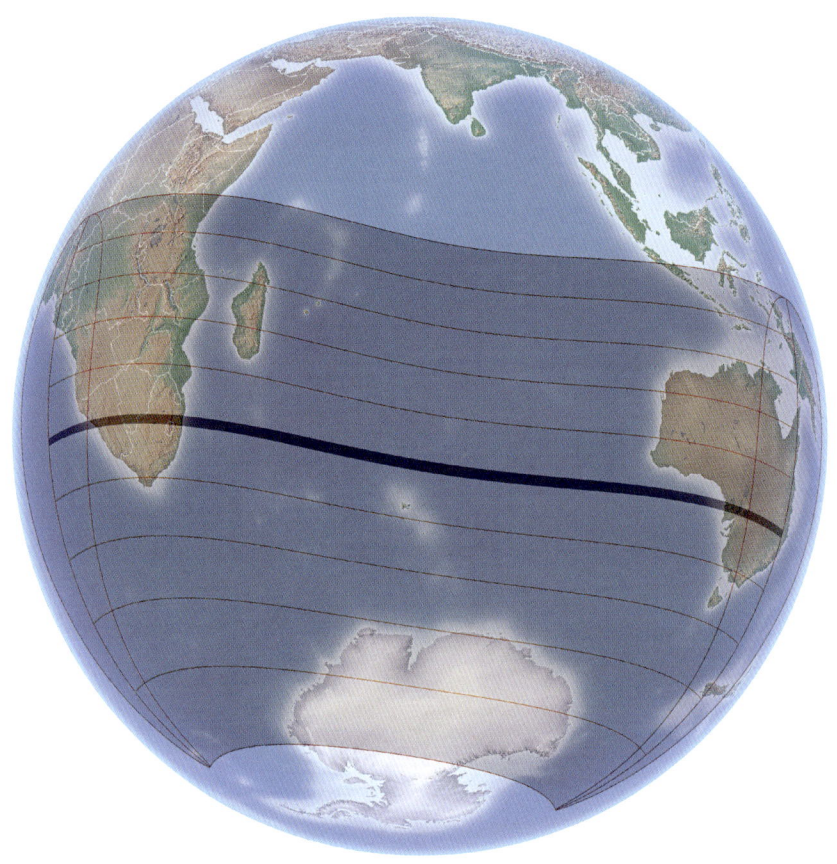

Saros series (number): 133 (46 of 72)		Magnitude: 1.05		
Greatest eclipse location: 43° 36.3' S, 071° 17.0' E		Duration: 3m 45s		
Partial begins (P1)	First totality begins (U1)	Greatest eclipse	Last totality ends (U4)	Partial ends (P4)
04:16:29 UT	05:14:07 UT	06:50:10 UT	08:26:11 UT	09:23:48 UT

Path of totality touches Namibia, Botswana, South Africa, Lesotho and Australia.

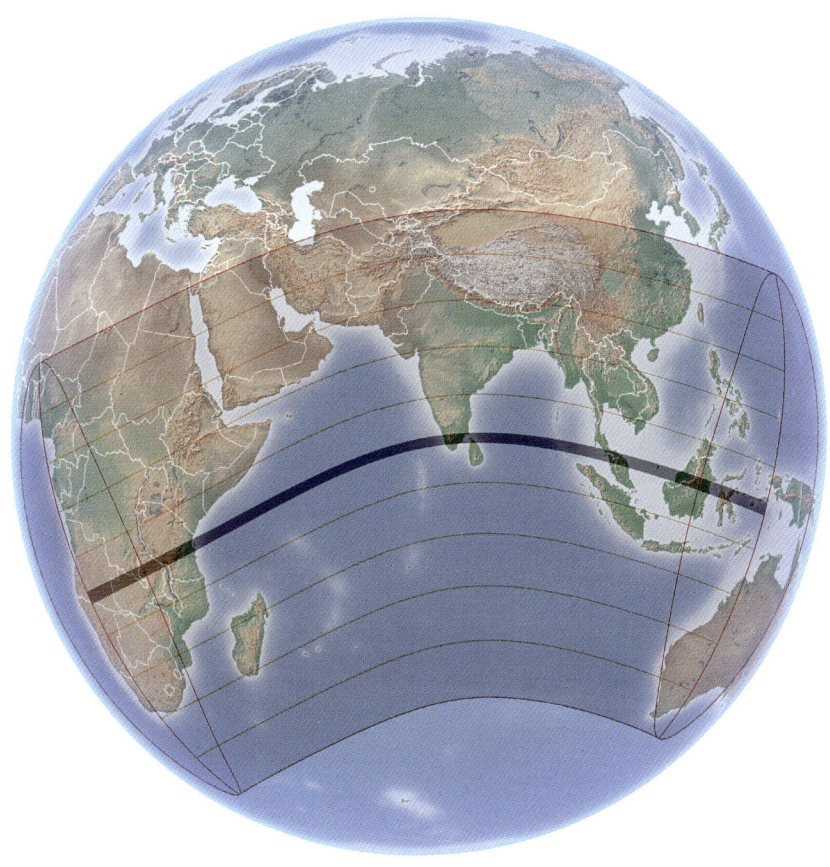

Saros series (number): 138 (32 of 70)			Magnitude: 0.96	
Greatest eclipse location: 08° 54.9' N, 071° 46.4' E			Duration: 5m 25s	
Partial begins (P1)	First annularity begins (U1)	Greatest eclipse	Last annularity ends (U4)	Partial ends (P4)
04:13:59 UT	05:18:28 UT	07:14:40 UT	09:10:55 UT	10:15:26 UT

Path of annularity touches Angola, Zambia, Democratic Republic of the Congo, Malawi, Tanzania, India, Sri Lanka, Thailand, Malaysia and Indonesia.

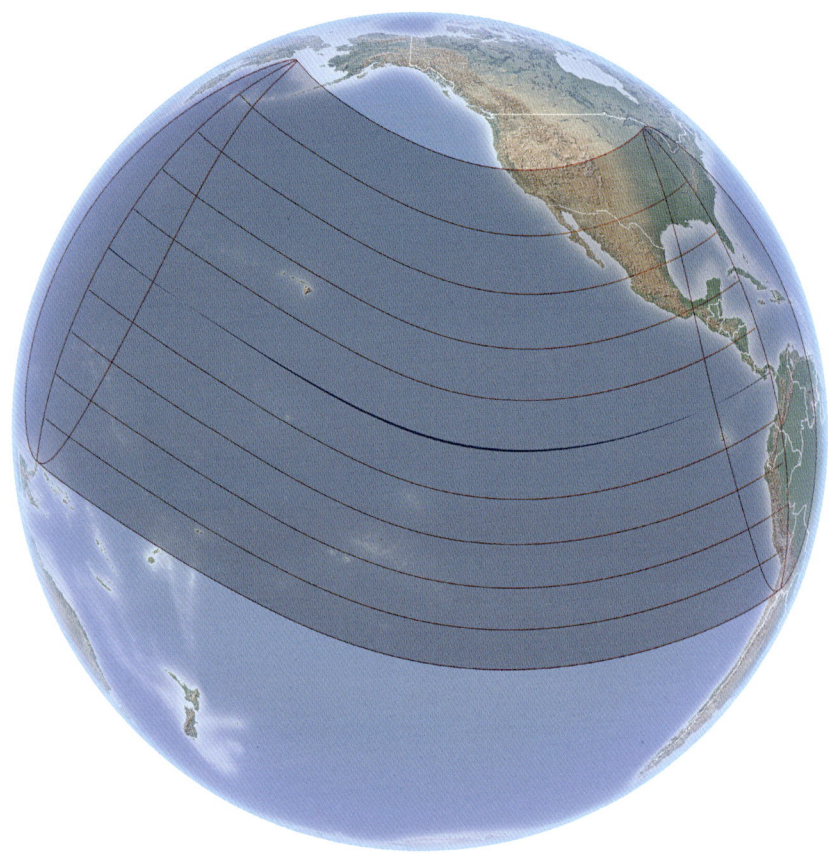

Saros series (number): 143 (24 of 72)			Magnitude: 1.01	
Greatest eclipse location: 00° 37.4' S, 137° 35.3' W			Duration (total): 1m 08s	
Partial begins (P1)	First annularity begins (U1)	Greatest eclipse	Last annularity ends (U4)	Partial ends (P4)
18:22:59 UT	19:23:38 UT	21:06:03 UT	22:48:28 UT	23:49:04 UT

Path of annularity touches Panama (annular).

Note: This hybrid eclipse begins and ends its central portion as an annular eclipse, but it does become a total eclipse over the Pacific Ocean.

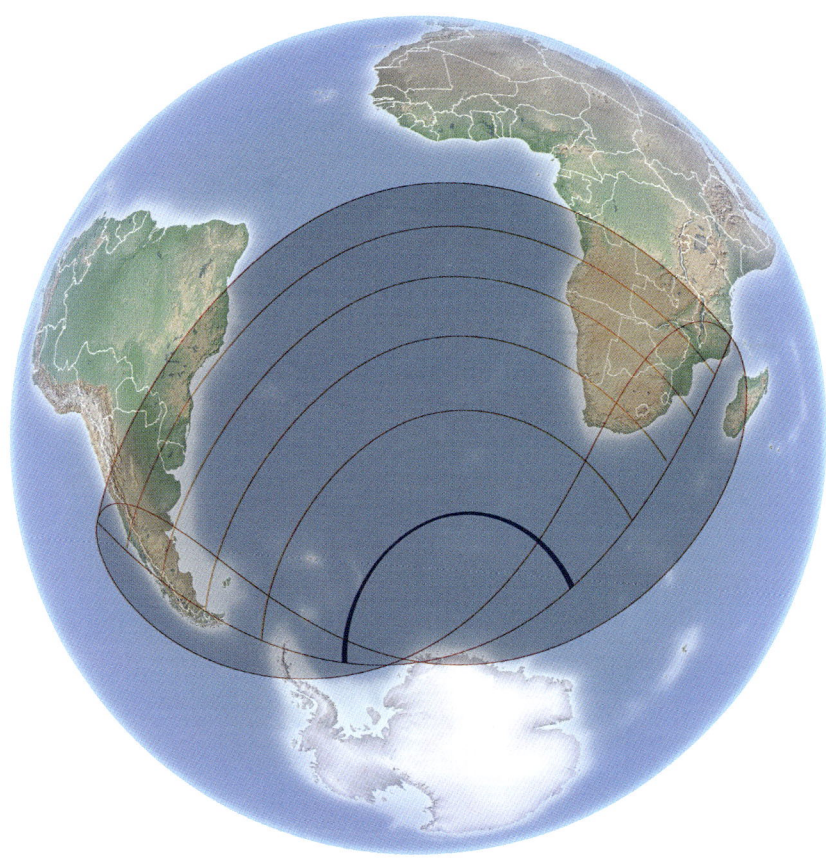

Saros series (number): 148 (22 of 75)			Magnitude: 0.99	
Greatest eclipse location: 51° 17.3' S, 007° 00.5' W			Duration: 22s	
Partial begins (P1)	First annularity begins (U1)	Greatest eclipse	Last annularity ends (U4)	Partial ends (P4)
11:09:40 UT	12:47:03 UT	13:25:15 UT	14:03:44 UT	15:41:04 UT

Path of annularity touches the south Atlantic Ocean.

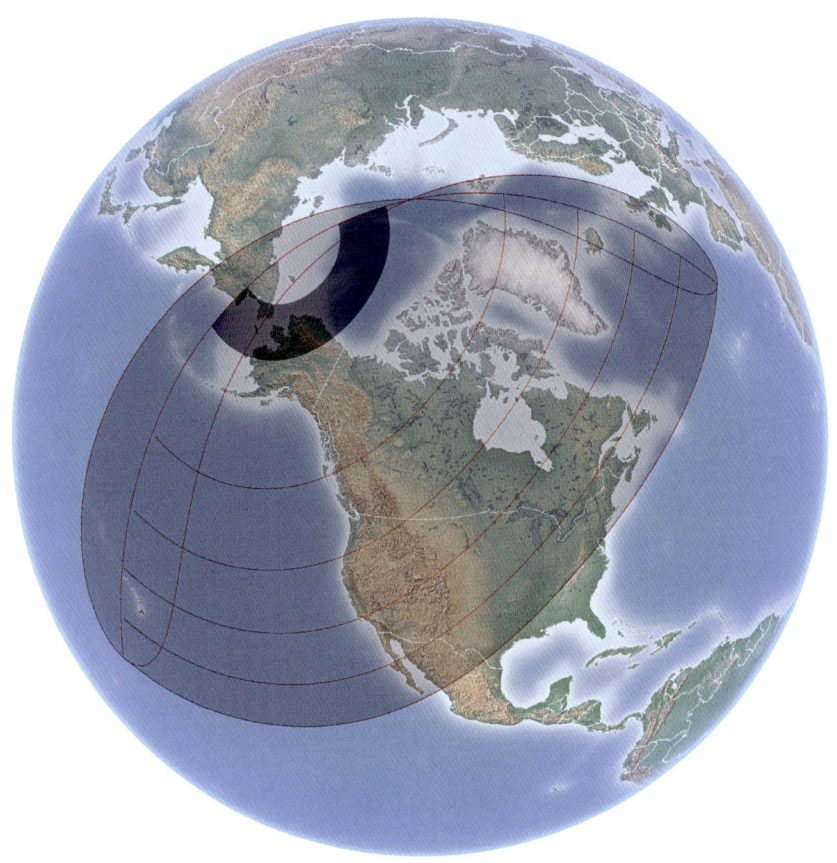

Saros series (number): 120 (62 of 71)		Magnitude: 1.05		
Greatest eclipse location: 71° 18.9' N, 155° 45.7' W		Duration: 2m 37s		
Partial begins (P1)	First totality begins (U1)	Greatest eclipse	Last totality ends (U4)	Partial ends (P4)
15:59:17 UT	17:35:34 UT	18:01:07 UT	18:26:17 UT	20:02:43 UT

Path of totality touches USA (Alaska) and Russia.

Saros series (number): 130 (53 of 73)			Magnitude: 1.05	
Greatest eclipse location: 16° 03.3' N, 022° 16.1' E			Duration: 4m 9s	
Partial begins (P1)	First totality begins (U1)	Greatest eclipse	Last totality ends (U4)	Partial ends (P4)
07:39:44 UT	08:36:42 UT	10:17:17 UT	11:57:47 UT	12:54:42 UT

Path of totality touches Benin, Nigeria, Cameroon, Chad, Sudan, Egypt, Saudi Arabia, Kuwait, Iran, Afghanistan, Pakistan, India and China.

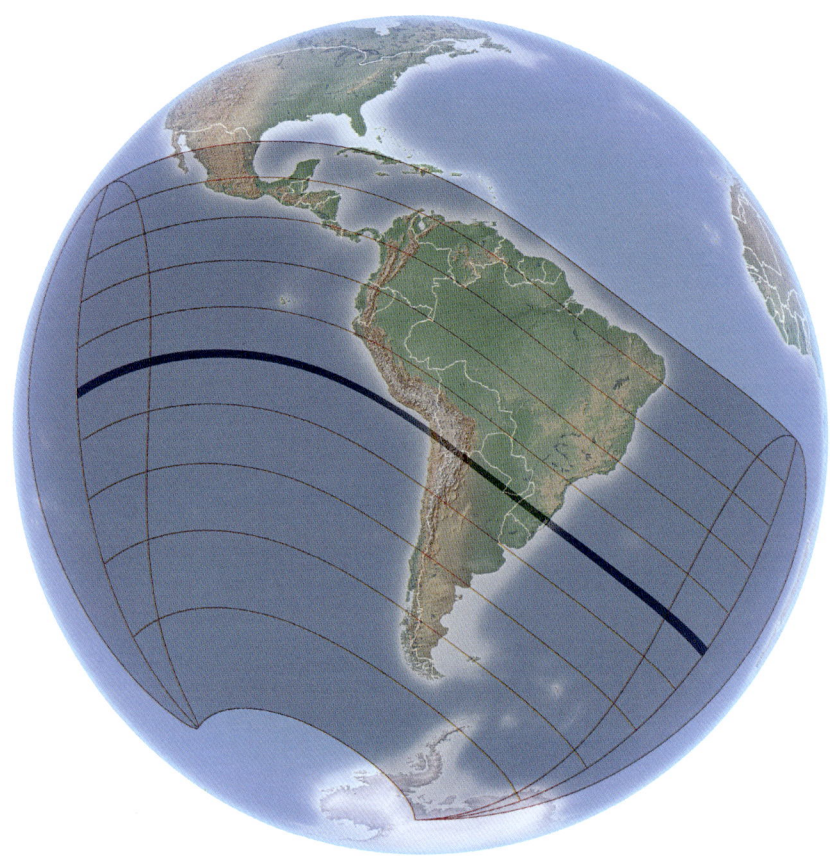

Saros series (number): 135 (40 of 71)			Magnitude: 0.97	
Greatest eclipse location: 18° 14.1' S, 072° 33.8' W			Duration: 2m 57s	
Partial begins (P1)	First annularity begins (U1)	Greatest eclipse	Last annularity ends (U4)	Partial ends (P4)
13:26:24 UT	14:31:53 UT	16:17:58 UT	18:03:56 UT	19:09:31 UT

Path of annularity touches Chile, Bolivia, Argentina, Paraguay and Brazil.

ANNULAR SOLAR ECLIPSE OF 9 MARCH 2035

Saros series (number): 140 (30 of 71)			Magnitude: 0.99	
Greatest eclipse location: 29° 02.9' S, 154° 54.2' W			Duration: 47s	
Partial begins (P1)	First annularity begins (U1)	Greatest eclipse	Last annularity ends (U4)	Partial ends (P4)
20:20:33 UT	21:24:48 UT	23:04:22 UT	00:44:03 UT	01:48:12 UT

Path of annularity touches New Zealand and French Polynesia.

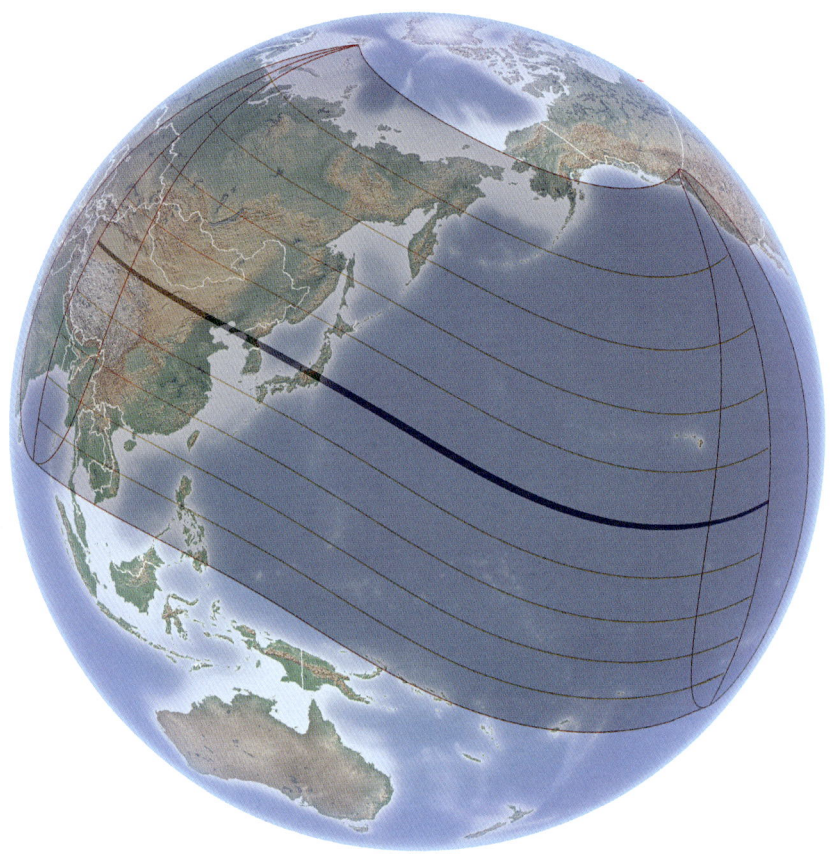

Saros series (number): 145 (23 of 77)			Magnitude: 1.03	
Greatest eclipse location: 29° 06.3′ N, 158° 04.8′ E			Duration: 2m 54s	
Partial begins (P1)	First totality begins (U1)	Greatest eclipse	Last totality ends (U4)	Partial ends (P4)
23:15:15 UT	00:15:34 UT	01:55:15 UT	03:35:02 UT	04:35:26 UT

Path of totality touches China, North Korea, South Korea and Japan.

OTHER ECLIPSES

This table lists non-central (penumbral and partial) eclipses until the end of 2036. As with the central eclipses, the date given is the UT date at greatest eclipse. Moderate and deep partial eclipses are particularly fine and have been highlighted in **bold**.

2026		
28 August	**Lunar (deep partial)**	**East Pacific, Americas, Europe, Africa**
2027		
20 February	Lunar (penumbral)	Americas, Europe, Africa, Asia
18 July	Lunar (penumbral)	East Africa, Asia, Australia, Pacific
17 August	Lunar (penumbral)	Pacific, Americas
2028		
12 January	Lunar (shallow partial)	Americas, Europe, Africa
6 July	Lunar (moderate partial)	Europe, Africa, Asia, Australia
2029		
14 January	**Solar (deep partial)**	**North America, Central America**
12 June	**Solar (moderate partial)**	**Arctic, Scandinavia, Alaska, Asia, Canada**
11 July	Solar (shallow partial)	Chile, Argentina
5 December	**Solar (deep partial)**	**Argentina, Chile, Antarctica**
2030		
15 June	**Lunar (moderate partial)**	**Europe, Africa, Asia, Australia**
9 December	Lunar (penumbral)	Americas, Europe, Africa, Asia
2031		
7 May	Lunar (penumbral)	Americas, Europe, Africa
5 June	Lunar (penumbral)	East Indies, Australia, Pacific
30 October	Lunar (penumbral)	Americas
2032		
3 November	**Solar (deep partial)**	**Asia**
2033		
23 September	**Solar (moderate partial)**	**South America, Antarctica**

2034		
3 April	Lunar (penumbral)	Europe, Africa, Asia, Australia
28 September	Lunar (shallow partial)	Americas, Europe, Africa
2035		
22 February	Lunar (penumbral)	East Asia, Pacific, Americas
19 August	Lunar (shallow partial)	Americas, Europe, Africa, Middle East
2036		
27 February	**Solar (moderate partial)**	**Antarctica, Australia, New Zealand**
23 July	Solar (shallow partial)	South Atlantic
21 August	**Solar (deep partial)**	**Alaska, Canada, Arctic, western Europe, north-west Africa**

Annular solar eclipse and the Statue of Liberty, New York.

GLOSSARY

angular diameter The apparent size of an object as seen from Earth, measured in degrees, arcminutes or arcseconds. Angular diameter depends on both the object's actual size and its distance from the observer.

angular size (of the Moon) The apparent diameter of the Moon as seen from Earth, averaging about 31 arcminutes or 0.52 degrees. This angular size varies slightly due to the Moon's elliptical orbit, appearing largest at lunar perigee and smallest at lunar apogee.

atoms The fundamental building blocks of matter, consisting of a nucleus surrounded by electrons. In astronomical contexts, atoms absorb and emit light at specific wavelengths, creating the spectral lines used to identify elements in stars and other celestial objects.

binary star systems A system of two stars gravitationally bound and orbiting around their common centre of mass. Binary systems are extremely common in the Galaxy, with many appearing as single stars to the naked eye due to their proximity or similar brightness.

chromosphere A region of the Sun's lower atmosphere, situated above the photosphere. The chromosphere is prominently revealed using a hydrogen-alpha solar telescope and appears as a thin pinkish layer during total solar eclipses

coronograph A specialized telescope designed to block the Sun's bright disk, allowing observation of the much fainter solar corona. Coronographs use an occulting disk to create an artificial eclipse, enabling study of coronal features and solar wind activity.

deferent circles In ancient geocentric astronomy, large circular orbits on which the centres of smaller epicycles were thought to move. This concept was part of the Ptolemaic system used to explain planetary motion before the heliocentric model was adopted.

draconic month The time period of 27.21 days for the Moon to return to the same position relative to the lunar nodes. This period is crucial for predicting eclipses, as solar and lunar eclipses can only occur when the Moon is near a node.

ecliptic An imaginary line tracing the apparent path of the Sun through the celestial sphere over one year. The ecliptic passes through the constellations of the zodiac, and the Moon, planets and asteroids always appear close to this path.

ecliptic plane The geometric plane containing Earth's orbit around the Sun. Most planets orbit close to this plane, and it serves as the fundamental reference plane for Solar System astronomy and coordinate systems.

epicycles In ancient astronomy, small circular orbits whose centres moved along larger deferent circles. This complex system was used in the Ptolemaic model to explain the retrograde motion of planets before the heliocentric model simplified planetary motion.

filament (Sun) A dark, ribbon-like structure visible against the Sun's surface when viewed in hydrogen-alpha light. Filaments are prominences seen in silhouette against the solar disk, consisting of cooler plasma suspended in the corona by magnetic fields.

gravity waves Oscillations in a fluid medium where gravity acts as the restoring force, commonly occurring in Earth's atmosphere and oceans. In atmospheric science, gravity waves are generated by air flowing over mountains or by convective activity, and they can affect weather patterns and atmospheric circulation. Evidence suggests gravity waves can be induced by solar eclipses.

grazing lunar occultations Rare events where a star appears to repeatedly disappear and reappear behind the Moon's limb due to lunar mountains and valleys. Timing these occultations provides precise measurements of stellar positions and lunar topography.

greatest eclipse The moment during an eclipse when either the Moon covers the maximum amount of the Sun's disk, or the Moon is maximally submerged in the Earth's shadow. This point represents the deepest phase of the eclipse and occurs along the path of totality or maximum obscuration.

heliacal rising The first appearance of a star or planet in the dawn sky after a period of invisibility due to its proximity to the Sun. This phenomenon was historically important for calendar systems and navigation in ancient civilizations.

ionization The process by which atoms or molecules gain or lose electrons, creating charged particles called ions. In astronomy, ionization occurs in stellar atmospheres, nebulae and Earth's upper atmosphere due to high-energy radiation.

ionized layers of Earth's atmosphere Regions of the upper atmosphere where solar radiation has stripped electrons from atoms, creating charged particles. These layers, including the ionosphere, reflect radio waves and can affect satellite communications and GPS signals.

Kuiper Belt A region of the Solar System beyond the orbit of Neptune, extending from about 30 Astronomical Units (AU) to 50 AU, where one AU is the average distance between the Earth and the Sun. Pluto is the most famous member of the Kuiper Belt, which is also home to many other small worlds composed of ice and rock.

Kuiper belt objects Small celestial bodies orbiting the Sun in the Kuiper Belt region beyond Neptune. These are remnants from the Solar System's formation and include dwarf planets such as Pluto and thousands of smaller icy bodies.

libration The apparent slight wobbling motion of the Moon that allows observers on Earth to see slightly more than half of the lunar surface over time. Libration occurs due to the Moon's elliptical orbit and tilted axis, enabling about 59 per cent of the Moon's surface to be seen from Earth.

lunar nodes The two points where the Moon's orbit intersects the ecliptic plane. The ascending node is where the Moon moves north of the ecliptic, while the descending node is where it moves south. Eclipses occur when the Sun, Earth and Moon align near these nodes.

lunar phases The changing appearance of the Moon as seen from Earth due to the varying angles between the Sun, Earth and Moon. The cycle includes New Moon, waxing crescent, first quarter, waxing gibbous, Full Moon, waning gibbous, third quarter and waning crescent. Solar eclipses always occur at New Moon, and lunar eclipses always occur at Full Moon.

Mare Crisium The 'Sea of Crises', a large lunar mare (dark basaltic plain) located on the eastern edge of the Moon's near side. This circular feature is about 178 km (111 miles) in diameter and is easily visible to the naked eye as a dark patch.

molecule A group of atoms bonded together by chemical forces. In astronomy, molecules are detected in interstellar space, planetary atmospheres and comets through their distinctive spectral signatures, providing information about cosmic chemistry and conditions.

moonglow The faint illumination of the sky surrounding the lunar disk, particularly visible during a bright phase such as a gibbous Moon or Full Moon.

obscuration The fraction of the Sun's disk covered by the Moon during a partial solar eclipse, typically expressed as a percentage. Maximum obscuration varies by location and can range from a small fraction to nearly complete coverage outside the path of totality.

occultation Event in which one celestial object passes in front of another, temporarily blocking it from view. Lunar occultations of stars and planets are commonly observed, providing precise timing measurements and information about both objects involved.

occulting object The celestial body that passes in front of and blocks another object during an occultation. The Moon is the most common occulting object as seen from Earth, regularly passing in front of stars and planets.

orrery A mechanical model of the Solar System that shows the relative positions and motions of planets and moons. Named after Charles Boyle, Earl of Orrery (1674–1731), these devices demonstrate orbital mechanics and celestial cycles through clockwork mechanisms.

ozone layer A region of Earth's stratosphere containing high concentrations of ozone (O_3) molecules, located approximately 15–35 km (9–22 miles) above the surface. This layer absorbs harmful ultraviolet radiation from the Sun, protecting life on Earth.

pascal The standard unit of pressure in the International System of Units, equal to one newton per square metre. In astronomy, pascals are used to measure atmospheric pressure on planets and the pressure within stellar interiors.

penumbral or shallow partial eclipses Lunar eclipses where the Moon passes through only the penumbra (outer shadow) of Earth, or just grazes the umbra, causing subtle darkening rather than dramatic colour changes. These events are often difficult to detect without careful observation.

photometry The measurement of the brightness and intensity of electromagnetic radiation from celestial objects. Photometry is fundamental to astronomy, allowing determination of stellar magnitudes, distances and physical properties of stars and other objects.

photon Particle of electromagnetic radiation that carries energy and momentum but has no mass. In astronomy, photons from stars and other celestial objects carry information about their temperature, composition and motion across vast distances.

quantum events Phenomena occurring at the atomic and subatomic scale where classical physics breaks down and quantum mechanics governs behaviour. In astronomy, quantum events are crucial for understanding stellar fusion, neutron star behaviour and the early universe.

radiosonde An instrument carried by a weather balloon to measure atmospheric conditions including temperature, humidity and pressure at various altitudes. These devices provide data essential for weather forecasting and atmospheric research.

Saros series A cycle of approximately 18 years, 11 days and 8 hours after which similar eclipses repeat due to the alignment of lunar orbital periods. Each Saros series contains dozens of eclipses occurring over many centuries with gradually shifting paths.

sidereal motion The apparent movement of distant stars as caused by Earth's rotation. Sidereal time is based on Earth's rotation relative to the stars rather than the Sun, making it essential for astronomical observations. One sidereal day is 23 hours, 56 minutes and 4 seconds long, a few minutes shorter than a civil day (or average solar day) of 24 hours.

solar corona The extended atmosphere of the Sun, visible by eye briefly during the totality of a total solar eclipse. The solar wind comprises material escaping the corona into the Solar System, and its temperature can exceed 1 million °C (1.8 million °F).

spectra The distribution of electromagnetic radiation from an object across different wavelengths or frequencies. Spectra reveal information about temperature, composition, motion and magnetic fields through characteristic absorption and emission lines. Spectra are produced using a spectrograph, which splits light into its component colours for analysis.

stellar magnitude A logarithmic scale measuring the brightness of stars and other celestial objects. The system originated with ancient Greek astronomers, where magnitude 1 represents the brightest stars and magnitude 6 represents the faintest visible to the naked eye. Each magnitude represents a difference of about 2.5x, such that magnitude 6 is 100x fainter than magnitude 0.

stratosphere A layer of Earth's atmosphere extending from about 20 to 50 km (12 to 31 miles) above the surface, characterized by increasing temperature with altitude. The stratosphere contains the ozone layer and is where most weather balloons reach maximum altitude.

sublunar point The point on Earth's surface where the Moon appears directly overhead at any given moment. This point moves across Earth's surface as the planet rotates, and also drifts due to the Moon's own movement in its orbit. The sublunar point is the closest point on the Earth's surface to the Moon.

subtend To extend across a particular angle as seen from a specific point. In astronomy, objects subtend angular sizes depending on their actual size and distance from the observer.

sunspot A dark feature visible on the Sun's photosphere with a white-light filter. A sunspot is a relatively cool region of the photosphere where convection of heat is restricted by intense magnetic stress, varying in number according to the solar cycle.

Surveyor 3 An unmanned lunar lander that touched down in Oceans Procellarum (the 'Ocean of Storms') in April 1967, part of NASA's Surveyor program. The mission demonstrated soft-landing techniques and was later visited by Apollo 12 astronauts who retrieved parts for analysis.

synodic month The time period of 29.53 days between successive New Moons or Full Moons, representing the complete cycle of lunar phases. This period forms the basis for many calendar systems and differs from the sidereal month due to Earth's orbital motion.

Total Electron Content (TEC) measurements Quantitative assessments of the total number of electrons in a column of Earth's atmosphere, typically measured between satellites and ground receivers. TEC measurements are crucial for GPS accuracy and space weather monitoring.

troposphere The lowest layer of Earth's atmosphere, extending from the surface to approximately 20 km (12 miles) in altitude. This layer contains most of Earth's weather phenomena and nearly all water vapour, with temperature generally decreasing with altitude.

umbral region The inner, darker part of a shadow cast by an opaque object. During solar eclipses, observers in the Moon's umbra experience totality, while the umbral region during lunar eclipses appears dark red due to refracted sunlight.

UT Universal Time, a time standard based on Earth's rotation relative to distant stars. UT serves as the basis for civil time worldwide and is essential for coordinating astronomical observations across different time zones.

yaw attitudes The orientation of a spacecraft or satellite around its vertical axis, representing rotation left or right from its nominal pointing direction. Yaw control is crucial for maintaining proper positioning of solar panels and communication equipment.

INDEX

ACKNOWLEDGEMENTS

Alamy Stock Photo: History & Art Collection 20, Hugo Martin 10, North Wind Picture Archives 18, Science History Images 22, The Reading Room 92, World History Archive 7, 12; **Alex Pietrow:** 83; **DCATT Team, MSX Project, BMDO (Ballistic Missile Defense Organization of the US DoD), IRSA (Infrared Science Archive), Tom Kerss:** 28; **Dylan O'Donnell, public domain via esa.int:** 27, 37; **©ESA:** B. Sicardy (LESIA, Observatoire de Paris, France), P. Tanga (Observatoire de la Côte d'Azur, Nice, France), A. Carbognani (Osservatorio Astronomico Valle d'Aosta, Italy), R. Leiva (LESIA, Observatoire de Paris) 42 (CC BY-SA 3.0 IGO) 77, 94, CESAR/Wouter van Reeven (CC BY-SA 3.0 IGO) 101bl, J. Spencer (Lowell Observatory)/NASA 39a, Proba-3/ ASPIICS/WOW algorithm (CC BY-SA 3.0 IGO) 34r; **ESO:** L. Calçada/M. Kornmesser/Nick Risinger (skysurvey.org) 43, P. Horálek 103, 110, 116b, R. Lucchesi (CC BY 4.0) 100; **Firefly Aerospace:** 74l, 74r; **Getty Images:** China National Space Administration (CNSA) via CNS/AFP 48, SSPL 24r, Werner Forman/Universal Images Group 13; **Nationwide Eclipse Ballooning Project/Virginia Tech:** 47; **NASA:** 44a, 73r, adapted by Ryan French 35, 101a, Aubrey Gemignani 80, Bill Ingalls 109, 113, DSCOVR EPIC 51, enhanced Image by Kevin M. Gill (CC BY 3.0) based on images provided courtesy of NASA/JPL-Caltech/SwRI/MSSS 38–39, ESA/ adapted by Ryan French 33, Goddard Space Flight Center Scientific Visualization Studio 49, Jacques Descloitres, MODIS Rapid Response Team, GSFC 46, JPL-Caltech/ASU/ MSSS/SSI 38a, JSC 29, 45, Joel Kowsky 36, Johns Hopkins University Applied Physics Laboratory/Southwest Research Institute 44b, Keegan Barber/adapted by Ryan French 98, Surveyor 3, R. D. Sampson (ECSU)/adapted by Tom Kerss 73l; **NASA/Dan Seaton/Matthew West:** 32b; **NOAA:** 34l; **NOIRLab:** N.A. Sharp/KPNO/NOIRLab/NSO/NSF/AURA (CC BY 4.0) 26b; **Peisang Tsai, Research Aviation Facility, Earth Observing Laboratory, Broomfield, CO.:** 32a; **Pexels:** Allan Carvalho 111, Max Ravier 66, Roberto Nickson 68a; **Pixabay:** KBOutdoors 65, 69; **Ryan French:** 81, 82a, 82b, 84a, 102; **Science Photo Library:** Royal Astronomical Society 24l, 25; **Courtesy of SwRI, Citizen CATE 2024/ Ritesh Patel/Dan Seaton:** 31; **Tom Kerss:** 4, 11, 67, 68b, 70, 71, 72, 75, 76, 105, 106, 107, 108a, 108b, 112, 114a, 114c, 114b, 115a, 115b, 116a, 117, 118–122, 123, 125–150, back cover; **©2025 UCAR:** 99; **United States Air Force (CC0):** 50; **sourced via Wikimedia Commons:** adapted by Ryan French from Michael S Adler (CC BY-SA 4.0) 97, Anthony Quintano 152, Archives des missions scientifiques et littéraires (CC0) 26a, E. A. Rodrigues (CC0) 14l, 14r, Frank Vincentz (CC0) 9, Harishnalrah (CC BY-SA 4.0) 90, Julius Söhn (CC0) 78, Kevin Gill (CC BY 2.0) 84–85, Kgbo (CC BY-SA 4.0) 21r, KnightFallV (CC BY-SA 4.0) 41, Luc Viatour (CC BY-SA 3.0) 87 Madeleine Slierstaart (CC BY-SA 3.0) 19, Michael S Adler (CC BY-SA 4.0) 3, Ranssom (CC BY-SA 4.0) 16-17b, skua47 (CC BY 2.0) 101br, Stephan Rahn (CC0) 89l, Steve Elliott (CC BY-SA 2.0) 89r, Tilemahos Efthimiadis from Athens, Greece (CC BY 2.0) 21l, Tomruen (CC BY-SA 4.0) 93, Total Eclipses of the Sun (1894), Mabel Loomis Todd, José Joaquín de Ferrer (illustration, 1806) (CC0) 86, Vladoubi-doOo, (CC BY-SA 3.0) 15, Wilhelm Meyer (CC0) 16a.